MEMS Pressure Sensors: Fabrication and Process Optimization

Parvej Ahmad Alvi

MEMS Pressure Sensors: Fabrication and Process Optimization

IFSA International Frequency Sensor Association Publishing

Parvej Ahmad Alvi
MEMS Pressure Sensors: Fabrication and Process Optimization

ISBN-10: 84-616-2207-3
ISBN-13: 978-84-616-2207-8
BN-20121212-XX
BIC: TJFD

Dedicated to my Parents

and my family

Acknowledgement

My Lord, most gracious & merciful, beginning his name

First of all I am highly indebted and thankful, the lord of universe, the most merciful and benevolent to all, who taught men that which he knew not and bestowed the men with epistemology and gave him the potential to ameliorate the same. It is His blessing, which inspired and enabled me to complete this book.

*I wish to express my profound gratitude, **Dr. K. M. Lal,** Professor and Former Chairman, Department of Applied Physics, Faculty of Engineering and Technology, Aligarh Muslim University, Aligarh, India for his excellent sprit, constant encouragement, illuminative and precious guidance.*

*I am also greatly indebted to **Dr. Jamil Akhtar,** Dy. Director and Scientist-G, Central Electronics Engineering Research Institute (CSIR), Pilani, Rajasthan, India for his excellent supervision, encouragement and valuable suggestions during my entire research work.*

*My Special thanks go to **Dr. S. Alim H. Naqvi** (Professor and Former Chairman, Department of Applied Physics, A.M.U., Aligarh), **Dr. Mohd. Jawaid Siddiqui** (Department of Electronics Engineering, A.M.U., Aligarh).*

Last but not least, I wish to a lot of thanks my parents and my wife (Zeba) and other family members for their constant affection, love and encouragement, which played a great role in the completion of my work.

Parvez Ahmad Alvi

Content

Preface

For the great progress of MEMS (Micro-Electro-Mechanical Systems) in recent years, there are at least four kinds of processing methods including silicon bulk micro-machining, silicon surface micro-machining, LIGA and CMOS (complementary metal-oxide-semiconductor) process to fabricate the micro-sensors at present. Among these technologies, CMOS process for micro-sensors has the advantages of the maturity in IC (integrated circuit) foundry, the sub-micrometer spatial resolution of device fabrication and the functionality of on-chip circuitry. CMOS layers have been successfully used as the mechanical structures or the sensing elements of accelerometers, thermal sensors, magnetic sensors and pressure sensors. A group of Scientist has also designed a magnetic Hall sensor by the standard SPDM (single-polysilicon-double-metal) CMOS foundry service provided by the Chip Implementation Center (CIC), Taiwan, without post-processing. However, it caused some difficulties in achieving the design and the fabrication of the CMOS mechanical sensors due to the violations of some electric rules provided by the CMOS foundry line and the uncertainty of post-processing used to form the special geometry of sensor chip.

However, recently a novel technique (Lateral front side etching technique) has been utilized to fabricate MEMS based micro pressure sensors. Front side etching technique is the most advanced and recently developed MEMS technology. It takes the advantages of both the bulk and surface micromachining technologies.Micro-electro-mechanical systems (MEMS) are Freescale's enabling technology for acceleration and pressure sensors. MEMS based sensor products provide an interface that can sense, process and/or control the surrounding environment.

Freescale's MEMS-based sensors are a class of devices that builds very small electrical and mechanical components on a single chip. MEMS-based sensors are a crucial component in automotive electronics, medical equipment, hard disk drives, computer peripherals, wireless devices and smart portable electronics such as cell phones. These sensors began in the automotive industry especially for crash detection in airbag systems. Throughout the 1990s to today, the airbag sensor market has proved to be a huge success using MEMS technology. MEMS-based sensors are now becoming pervasive in everything from inkjet cartridges to cell phones. Every major market has now embraced the technology.

Fabrication of thin membranes has been an important aspect in common micromechanical devices owing to its numerous industrial applications. The pressure sensing technology that provides a multiple-measurement and multiple-range capability is also based on a thin diaphragm or membrane fabrication process. This book describes the experimental details of the fabrication of a thin membrane over a conical V-shaped cavity using front side lateral etching technology and the results obtained are discussed.

In the reported work, front side lateral etching technology has been studied. This study proposes a novel front side lateral etching fabrication process for silicon based piezoresistive pressure sensor. As far as the fabrication process is concerned, this technique successfully accomplished a front side etching process laterally to replace the conventional back-side bulk micromachining. This novel structure pressure sensor can achieve the distinguishing features of the chip size reduction and fabrication costs degradation.

The study presented in this book is divided into eight chapters. Work reported in the chapters 2, 3, 4, 5, 6 and 7 is the original contribution of the author. Each chapter is started with an introduction and ending with references as follows:

Chapter 1: Introduction

Chapter 2: Micro-fabrication technologies

Chapter 3: Thin Film Materials

Chapter 4: Anisotropic Etching

Chapter 5: Fundamental Theory and Design of Pressure Sensor

Chapter 6: Fabrication of Pressure Sensor using Front-Side-Lateral Etching Technology

Chapter 7: In-process Observations and Discussions

Chapter 8: Summary and Conclusion

14

Chapter 1 describes various types of micromachining technologies to fabricate microstructures. This chapter also describes transduction mechanism and micro-machined pressure sensors with the details of reported macroscopic devices.

Chapter 2 is based on micro-fabrication technologies (e.g., submicron photolithography, thermal oxidation, thin film deposition and chemical etching etc.) to construct the three dimensional micro-electro-mechanical (MEMS) devices such as the fabricated device- absolute micro pressure sensor. The fabrication of the pressure sensor exploits the reproducibility and repeatability of these semiconductor technologies.

Chapter 3 entitled "Thin Film Materials" is focused on thin film materials like silicon dioxide, silicon nitride and especially on polysilicon thin film. Mechanical and electrical properties of polysilicon thin film deposited by LPCVD have been studied and discussed. Control of the electrical conductivity of the polysilicon layer is carried out using doping of compatible specie such as boron for p-type. The variation in sheet resistivity of polysilicon layer with increasing doping temperature is studied for the doping of boron and phosphorous in polysilicon by diffusion technique taking into account the underneath layer over which the polysilicon film is deposited. The variation in sheet resistivity with doping concentration of boron and phosphorous is also reported in this chapter. Thermally treated polysilicon film has been analyzed for its topological details using AFM in contact mode under ambient temperature and pressure. The grain size of the polysilicon film has been observed invariant with varying temperatures during boron doping.

The fundamental processing sequence common to most MEMS devices calls for the deposition of a thin film onto a sacrificial layer, which is subsequently etched away. Manufacturers commonly use polysilicon as a sacrificial layer, as it is being used in the present work, on $SiO_2/Si3N_4$ because of the excellent etching selectivity in the hydrofluoric acid (HF) between the two films. This approach exemplifies the importance of advanced micro-fabrication technology in creating advanced micro-electro-mechanical (MEMS) devices such as the reported device in the present work.

Chapter 4 on "Anisotropic Etching" starts with the introduction of anisotropic etching giving the details of experimental methodology. The experimental data during wet anisotropic etching using aqueous

KOH solution have been recorded in this chapter. Surface roughness of Si (100) surface during KOH etching has been studied. After studying the AFM images of the etched surfaces, it is concluded that surface quality improves (roughness decreases) with increasing KOH concentration. From the experimental data recorded, faster etching rate with smooth silicon surface has been observed at higher temperatures in low KOH concentration solutions. Higher KOH concentrations improve etched surface finish. Etch rate for a particular concentration increases with temperature.

Chapter 5 entitled as "Fundamental Theory and Design of Pressure Sensor" gives the description of theory and design layout of the fabricated pressure sensor. In order to do stress analysis for uniformly loaded square thin diaphragm or membrane under the effect of pressure load, we have dealt with energy method analysis and it is concluded that thin diaphragm will offer maximum stress and hence maximum sensitivity of the device. The variations in maximum pressures as a function of membrane area for a clamped membrane of composite layer has been shown, where the membrane thickness is used as a parameter. The design part largely depends on the simulated results of the stress profile on a strained diaphragm, which helps in the placements of resistors and also in the optimization of their shapes.

Chapter 6 entitled as "Fabrication of Micro Pressure Sensor (using Front - Side- Lateral Etching Technology)" contains the study on front side etching technology and its application in the fabrication of absolute micro pressure sensor in detail. The reported technology (i.e. front side lateral etching technology) is different from the front side normal etching technology in the aspect of operation, easy handling and fabrication processing. The micro pressure sensor has been fabricated by two processes; the sequences of both processes have been discussed step by step along with their flow charts giving the complete details.

Chapter 7 on "In-Process Observations and Discussion" shows and analyses the corresponding photographs observed after each step during fabrication of absolute micro pressure sensor for both the fabrication processes.

Chapter 8, the last chapter of this book, presents "Summary and Conclusions" of the entire work.

A process for the fabrication of a vacuum sealed cavity in (100) silicon with the piezoresistors configured in half Wheatstone bridge has been developed by integrating conventional microelectronic processes with MEMS technology using front side etching technique. The sealing of the cavity provides a way to set the desired range of the pressures to be measured. The technology is compatible to fabricate tiny cavities of micron size for a number of applications as bio sensors. The feasibility of the process has been shown by delineating polysilicon resistors of desired values over the cavity. The polysilicon resistors can be realized technologically by controlling the sheet resistivity of the polysilicon layer using doping of requisite species. Use of polysilicon as a sacrificial layer is an extra advantage while doing anisotropic etching of (100) silicon under the membrane.

The process has novelty in making thin membranes of external layers in silicon while maintaining advantages of bulk micromachining. Size of the cavity can be reduced to the limits of photolithography used in the process.

Hopefully, this book will be very beneficial to the students of MEMS and NEMS courses at under graduate and post graduate levels, as well as to technologists.

About the Author

Dr. Parvej Ahmad Alvi is an Assistant Professor in the Department of Physics, School of Physical Sciences, Banasthali University, India. He has got a PhD in Applied Physics (2009) from Department of Applied Physics, Faculty of Engineering & Technology, Aligarh Muslim University, Aligarh, India. His research area is MEMS, NEMS technologies, Opto-electronics, and Material Science. He has active collaboration with CEERI (CSIR) Pilani (India), Elettra Synchrotron (Italy), Aligarh Muslim University, Aligarh (India), and Royal University of Phnom-Penh (Cambodia). Dr. Ahmad Alvi has published many international research papers and articles in his field and four books. He is an editorial board member of International Journal of Optoelectronics Engineering (Scientific Academic Publishing, USA), Nanoscience and Nanotechnology (Scientific Academic Publishing, USA) and lifetime member of International Union of Crystallography.

Chapter 1

Introduction

Micro-electro-mechanical systems (MEMS) are the integration of the electronic circuits and the mechanical components through IC fabrication processes. The design rules and fabrication processes of the electronic circuits have been well established due to the development of semiconductor technology in the past four decades. However, the capability of fabricating mechanical components of MEMS devices has also been improved in the past fifteen years. At present, surface and bulk micromachining are two existing silicon fabrication technologies used to make mechanical components. Both of these two technologies are developed from the standard semiconductor fabrication processes. The flexibility in making the mechanical components of the MEMS devices is still limited by the fabrication processes, especially for the bulk micromachining technology. The MEMS devices fabricated through the bulk micromachining technology are constructed by several basic mechanical structures such as cavities, grooves, holes, membranes and suspensions. Due to the fabrication processes, most of the mechanical structures are considered to be located inside a rectangular unit for the (100) substrate.

However, the front-side etching technology is the most advanced and recently developed MEMS technology. Basically, it is the combination of both bulk and surface micro machining technologies. This technology is capable to control micrometer feature size machining in addition to being compatible with conventional microelectronics. A detail of the MEMS technology is presented in the following.

1.1. MEMS Technology

MEMS – Micro Electro Mechanical Systems also known as micro-system or micro-fabrication technology is 'miniaturization engineering', a multi-disciplinary approach to enable batch fabrication of three-dimensional mechanical structures, devices and systems – at least with one of the dimensions in microns or less. The emerging concept is based on the available micro-manufacturing options, material properties and the scaling laws referred to the application under investigation. As the name implies the micro establishes the dimensional scale, electro suggests either electricity or electronics or both and the mechanical implies 'moving' device components with degrees of freedom in translation, rotation, tilt or a combination of the either. Over the last decade, however the MEMS concept has grown to encompass other micro and sub-micron devices – with or without moving parts, which respond and measure micro or nano – level changes in physical quantities including thermal, magnetic, piezoelectric, optical and pressure variations.

The origin of MEMS is generally traced to R. P. Feynman's hypothesis [1] on miniaturization of devices and systems to the extent till physical laws and material properties impose no limits. The logical extension of the miniaturization derive in the field of 'micro-machines' or micro-systems with moving parts has come to be known as MEMS. The technology is based on the state-of-the-art integrated circuit (IC) fabrication techniques and methodologies and hence exhibits many advantages indigenous to IC technology. A few of those include cost reduction through batch fabrication, device-to- device consistency from advanced lithography and etching techniques, and general performance enhancement from dimensional down scaling, leading to size and weight reduction. In addition, by using materials such as silicon and fabrication techniques comparable with IC technology, MEMS mechanical components can be made monolithically integrated with electronics, producing a complete smart system–on–a chip that interacts with the physical world, performs electronic computations and communicates with other systems if necessary. These advanced characteristics have positioned MEMS to be a winning technology in many application arenas, a few of which are given in Table 1.1. Polysilicon piezoresistive pressure sensors, based on MEMS technology have been developed. The developed fabrication process is competent to provide nearly 80 chips of 4×4 mm^2 size on a two-inch diameter silicon (100) wafer with polished surfaces [2]. The application of the MEMS technology in the radio frequency regime referred to as RF MEMS.

MEMS have been in existence almost since the inception of microelectronics and IC technologies they are based upon. The discovery of piezoelectric effect in silicon and germanium in 1954 [11] is commonly cited as the stimulus for silicon based sensors and micro

22

machining [12, 13]. Silicon piezoresistors were bonded to metal diaphragms to create pressure sensors in the late 1950s. Even as early as the 1960s different techniques for bulk and surface micro machining were emerging.

Table 1.1. MEMS applications and devices.

Applications	Devices
Inertial Measurement	Accelerometers, gyroscopes [3]
Pressure Measurements	Pressure sensors for Automotive, Medical and Industrial Applications [4]
Micro – fluidics, Bio MEMS	Ink jet nozzles [5], mass flow sensors, micro droplet generators [6] and bio – lab chips [7]
Optics and communication	Digital micro mirrors, Optical switches and displays [8]
RF Communication	Switches, Inductors, Capacitors, Resonators and systems based on basic RF MEMS [9]
Others	Micro relays, disk heads and sensors [10]

The Resonant Gate Transistor of Nathanson and Wickstrom in 1965 [14] is widely recognized as one of the first applications of a micro mechanical device on a silicon substrate [15]. Traditionally, micro machining has been built largely upon IC technologies. The infrastructure that has developed for creating high quality single crystal silicon and thin film growth and patterning has served the smaller micromachining industry well.

The term "micromachining" usually refers to the fabrication of micromechanical structures with the aid of etching techniques to remove part of the substrate or a thin film. Silicon has excellent mechanical properties, making it an ideal material for machining. The micro machining is often divided into three categories: (i) Bulk micro-machining; (ii) Surface micro-machining, and (iii) High aspect ratio micro-machining, which will be presented here as such.

1.1.1. Bulk Micro-machining

Bulk micro-machining is often described as a subtractive process, where the bulk of the substrate (usually single crystal silicon) is etched, cut, or otherwise modified to make the desired structure. The substrates can be machined by numerous techniques including isotropic etching [16-19], anisotropic etching [20-24], electrochemical etching [15], spark machining [25], mechanical milling [26], ultrasonic milling [27], laser and laser-assisted etching [28, 29], and electro-discharge machining [30]. A second definition of bulk micro-machining would be the formation of a desired microstructure by utilizing the bulk of a substrate. In microelectronics terms these structures are usually relatively large. This second definition may be more appropriate, since it is inclusive of wafer bonding technology. The most widespread techniques for bulk micro-machining are wet anisotropic etching and wafer bonding.

Wet anisotropic etching is used on single crystal silicon wafers, where the etchant selectively attacks all crystalline planes faster than the {111} planes. In {100} silicon this can be used to create diaphragms, V-grooves, and the other structures. Diaphragm thickness can be controlled by using an electrochemical etch stop. Figure 1.1 illustrates the etching of a diaphragm in {100} silicon. In {110} silicon, {111} planes can intersect {110} planes at $90°$ angle, enabling the etching of deep vertical walls.

A wide variety of anisotropic etching solutions can be used, including ethylene diamene pyrocatechol (EDP) [31] and hydrazine [32]. Aqueous hydroxide solutions are also commonly used, including CsOH [33], KOH [20, 23, 24], NH_4OH [34], NaOH [35], and tetra-methyl ammonium hydroxide (TMAH) [36, 37]. Common figures of merit for etchants are (100): (111) etch rate selectivity, (100): (mask material) etch rate selectivity, and smoothness of etched surface. EDP and KOH are the most widely used and characterized etchants [15]. EDP has the advantage over KOH of better selectivity to the etch mask of SiO_2. However, KOH has superior (100): (111) etch rate selectivity. KOH contaminates silicon with potassium, but by proper chemical post cleaning much of the potassium can be removed [38]. Organic etchants such as EDP and TMAH generally do not pose ionic contamination issues.

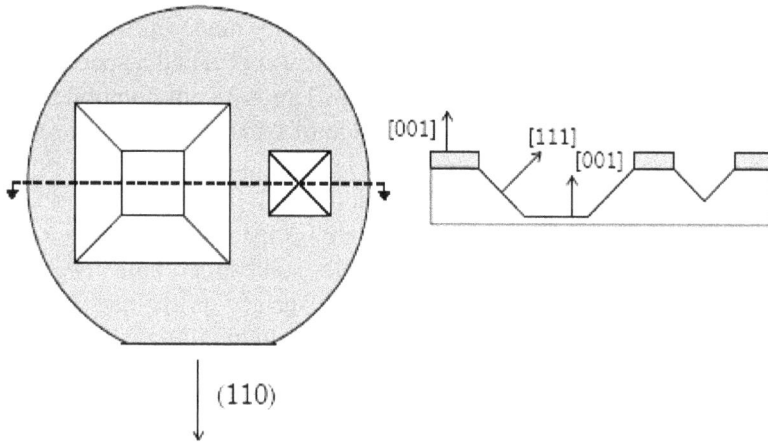

Figure 1.1. Anisotropic etching of {100} Silicon, cross section and plan view. Gray material is the etch mask, (e.g. SiO_2, Si_3N_4). The {111} sidewalls subtend 54.7° with the surface of the wafer. The etch pit on the right is self-terminated due to the intersection of {111} planes.

There are some problems with bulk micro-machining techniques: Bulk micro-machining involves extensive real estate consumption. In creating small membranes a large amount of Si - real estate is wasted and the resulting device becomes quite fragile. One major problem is the difficulty in etching convex corners. When etching rectangular convex corners, deformation of the edges occurs due to under cutting. This is an unwanted effect, especially in the fabrication of, for example, acceleration sensors, where total symmetry and perfect 90° convex corners on the proof mass are mandatory for good device prediction and specification. The undercutting is a function of etch time and thus directly related to the desired etch-depth.

1.1.2. Surface Micro-machining

In contrast to bulk micro-machining in which substrate material is removed by physical or chemical means, the Surface micromachining technique builds microstructures by adding materials layer by layer on top of the substrate; therefore it is often described as an additive technology. Typically, the desired microstructure is built by depositing and patterning thin films of structural (or mechanical) and sacrificial materials on the surface of the substrate.

Surface- micro-machined devices are typically made up of three types of components: (i) a sacrificial components (also called a spacer layer); (ii) a micro-structural component, and (iii) an insulator component. The sacrificial components are usually made of phosposilicate glass (PSG) or SiO$_2$ deposited on substrate by CVD techniques. These components in the form of thin films can be as long as 1 to 2000 µm and 0.1 to 5 µm thick. Both micro-structural and insulator components can be deposited in form of thin films. The most important mechanical materials for silicon technology based surface micromachining are polysilicon, silicon nitride, silicon dioxide, and aluminium. The etching rates for the sacrificial components must be much higher than those for the two other components.

Figure 1.2 illustrates the basic surface micromachining process. First a sacrificial layer is deposited and patterned (Fig. 1.2 (a), (b)), followed by a similar process for the mechanical layer (Fig. 1.2 (c), (d)). Finally, the sacrificial layer is etched to leave the free - standing structure (Fig. 1.2 (e)). An overview of surface micro-machining technique is given in Ref. [39]. The dimensions of these devices are quite different from those of bulk micro-machining.

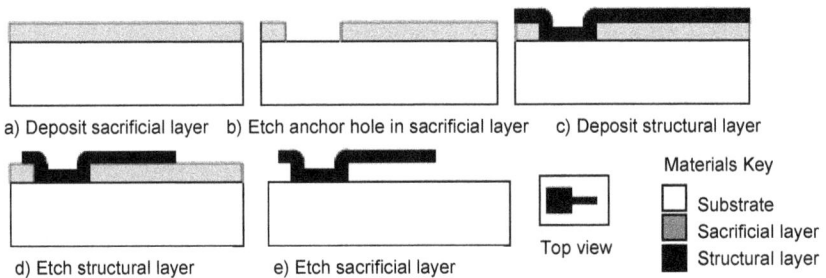

a) Deposit sacrificial layer   b) Etch anchor hole in sacrificial layer   c) Deposit structural layer

d) Etch structural layer   e) Etch sacrificial layer   Top view

Materials Key
Substrate
Sacrificial layer
Structural layer

Figure 1.2. Basic surface micro-machining process.

The suitable sacrificial layer depends upon the mechanical layer used, with the important factor the availability of an etchant that etches the sacrificial layer without significantly etching the mechanical layers or the substrate. A commonly used combination is an oxide sacrificial layer and a polysilicon or silicon nitride mechanical layer. Examples of these layer combinations can be seen in Table 1.2.

Table 1.2. Layer combinations and enchants for surface micromachining.

Sacrificial layers	Mechanical layers	Sacrificial Etchants
Oxides (PSG, LTO, etc.)	Polysilicon, Silicon nitride, Silicon carbide	HF
Oxides (PSG, LTO, etc.)	Aluminium	Pad etch, 73 % HF
Polysilicon	Silicon nitride	KOH
Polysilicon	Silicon dioxide	TMAH
Resist	Aluminium	Acetone-oxygen plasma

The principal advantages of the surface micromachining over the bulk micromachining are size and dimensional control. Figure 1.3 illustrates these ideas. Because of the nature of anisotropic etching, a bulk micro-machined diaphragm assembly must be at least the diameter of the diaphragm plus approximately 2 times the thickness of the wafer.

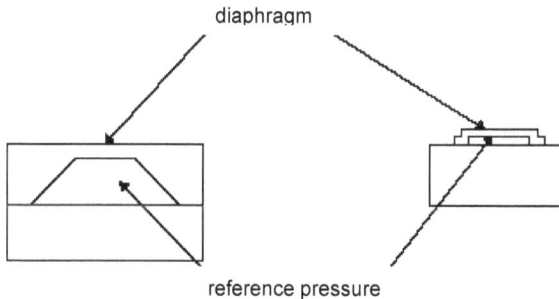

diaphragm

reference pressure

Figure 1.3. Size comparison of a bulk (left) and surface (right) micro-machined diaphragms. The surface micro-machined diaphragm has the potential of being smaller then the bulk micro-machined diaphragm.

Furthermore for a bulk micro-machined part, there can be the added complexity of aligning front side structures to the diaphragm (which is created by etching from the backside). Also, the dimensions of a bulk micro-machined diaphragm depend on wafer thickness, which is not always well controlled [15]. The surface micro-machined diaphragm assembly can be much smaller, approximately the diameter of the diaphragm itself. Front - to - backside alignment is unnecessary, and the dimensions are determined by the photolithography. However the mechanical properties of a deposited surface micro-machined diaphragm will, in general, not be as uniform and repeatable as a high quality, single crystal, and bulk micro-machined diaphragm.

1.1.3. High Aspect Ratio Micro-machining

High aspect ratio micro-machining (HARM) is similar to surface micromachining in that the desired structure is built upon the surface of the substrate. However, film thicknesses are typically on the order of 100 μm and vertical to lateral aspect ratios are on the order of 100:1. Furthermore, many HARM techniques are meant to be extended to molding processes so that relatively expensive micro-machined molds can be used to create inexpensive injection molded parts.

1.1.3.1. LIGA

A number of non-silicon based technologies have emerged in recent years. Probably the most famous of these is LIGA [40, 41]. This technology is a move away from the trend to make devices smaller and smaller. The strength of the LIGA process is the ability to make a high aspect ratio structures using an x-ray photoresist, such as poly-methyl-meth-acrylate (PMMA) [42, 43].

The term LIGA is an acronym for the German terms **LI**thographie, **G**alvanoformung, **A**bformung (Lithography, Electroplating, Molding in English). This technique was first reported by researchers at the Karlsruhe Nuclear Research Center in Karlsruhe, Germany [40, 44]. The words appearing in the term LIGA indeed represent the three major steps in the process, as outlined in Figure 1.4.

The LIGA process begins with deep x-ray lithography that sets the desired patterns on a thick film of photoresist. X-rays are used as the light source in photolithography because of their short wavelength, which provides higher penetration power into the photoresist materials. This high penetration power is necessary for high resolution in lithography, and for a high aspect ratio in the depth. The short wavelength of x-ray allows a line width of 0.2 μm and an aspect ratio of more than 100:1 to be achieved. The x-rays used in this process are provided by a synchrotron radiation source, which allows a high throughput because the high flux of collimated rays shortens the exposure time.

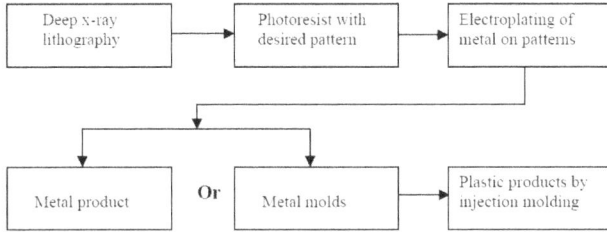

Figure 1.4. Major fabrication steps in the LIGA process.

The LIGA process outlined in Figure 1.4 may be demonstrated by a specific example as illustrated in Figure 1.5. The process begins by depositing a thick film of photoresist material on the surface of a substrate (Figure 1.5 (a)). The thick photoresist is exposed to synchrotron radiation (Figure 1.5 (a)) and developed (Figure 1.5 (b)).

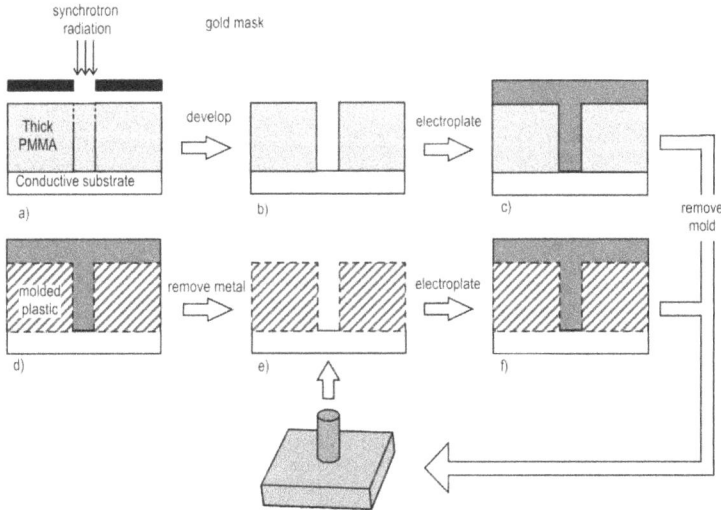

Figure 1.5. LIGA Processes: (a) Expose thick PMMA to synchrotron radiation, (b) Develop PMMA, (c) Electroplate metal into developed cavity; remove metal from PMMA mold. At this step finished metal part is achieved. (d and e) use metal part to create plastic mold, (f) Electroplate metal into plastic mold.

X-ray masks are typically composed of high atomic number elements, such as gold, in the opaque regions, and low atomic number materials, such s titanium or silicon nitride, in the transparent regions. The synchrotron radiation must typically have energy ≥ 1 GeV, which corresponds to the optimal energy for exposing PMMA film thickness of ≥ 100 μm [45].

Furthermore the exposures last several hours [42]. The assembly is then placed in an electroplating bath where a metal is deposited (Fig. 1.5(c)). After electroplating, the metal is then freed from the PMMA. This metal part can be used as it is. However to maximize cost benefits, the metal part can be used as a master to create plastic molds. The plastic molds, in turn, are used to fabricate more metal parts (Fig. 1.5 (d-f)). LIGA can produce very precise parts, but is limited because synchrotron sources are expensive, and geometric complexity is restricted. Much of the work to date has been done to develop the metal masters, although recently researchers at Karlsruhe have reported work with injection molding [46].

1.1.3.2. Other Techniques

Other high aspect ratio processes have been reported. One group of researchers used photosensitive polyamides 1.60 μm thick with nearly vertical sidewalls, while other groups have used photoresist [47, 48]. Several processes using deep dry etching into single crystal silicon substrate have been reported [49-52]. These alternative high aspect ratio processes are attractive because of lower cost, but may not be able to achieve the resolution and aspect ratios of LIGA.

1.2. Transduction Mechanism

Transduction is the means by which one form of energy is transformed into another. An example is the strain gauge, which transforms strain into a change of electrical resistance. Some methods of transduction are based on the fundamental physical laws, such as electrostatic attraction. Other methods, however, exploit peculiar materials properties. For example, a strain gauge takes advantage of the observed phenomenon of piezoresistance: that is, the change in resistance of a conductor due to an externally applied strain. This effect is discussed in the next section.

1.2.1. Piezoresistance

Piezoresistance is the property that a material changes its resistivity due to an applied strain. The resistivity change is generally linear with strain. While piezoresistivity is present in most metals, the effect in semiconductors is up to two orders of magnitude stronger [53]. Piezoresistivity was first reported in Si and Ge in 1954 [54]. The large effect in silicon and germanium is due to electronic band deformation and redistribution of carriers within the various conduction and valance bands [53]. Piezoresistance is useful whenever a direct strain is to be measured, or when a physical variable can be related to strain. It has been applied to pressure sensors and accelerometers [55, 56]. These two devices typically have diaphragms and proof masses deflect, respectively, which apply strain to piezoresistors. Two such sensors are shown schematically in Figure 1.6.

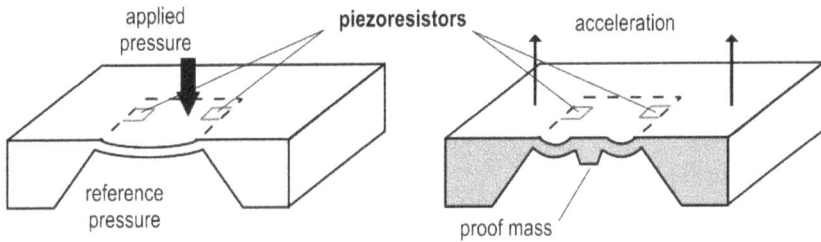

Figure 1.6. Cross section schematic of piezoresistive pressure sensor (left), and accelerometer (right).

1.2.2. Piezoelectricity

The piezoelectric effect was first discovered in materials exhibiting the pyroelectric effect. When a piezoelectric material is stressed, an electric flux density is generated in the material. In general, the stress tensor σ_{ij} is a second order tensor with nine terms. The electric field rising from an applied stress is called a polarization, Pi (C/m^2), which is a first order, three-term tensor. Hence the piezoelectric module, d_{ijk}, are described by a third order tensor with 27 terms, and the relationship can be expressed as

$$P_i = d_{ijk}\sigma_{ij}$$

Conversely, the material itself can be deformed if an electric field is placed across it:

$$\varepsilon_{jk} = d_{ijk}E_l,$$

where ε_{jk} is the resulting strain from an applied electric field E_l [V/m].

Piezoelectricity occurs in materials whose crystalline structure lacks a center of symmetry. In crystalline materials, of the 32 crystalline classes, 21 classes lack a center of symmetry, and of those 21, 20 can show piezoelectricity. All piezoelectric materials are pyroelectric; that is, upon application of a temperature gradient, an electric field is generated. Also, it is noted that some polymers, such as polyvinyledine fluoride exhibit the pyroelectric effect. In practice, piezoelectric materials have been widely used in sensor applications, specifically in sonic and ultrasonic applications, such as sonar, imaging, range finding, microphony, etc. For both sensor and actuator applications, large piezoelectric coefficients are desired. Large polarizations and large deflections, respectively, are achieved from large coefficients.

1.2.3. Bimorphs

Bimorphs are actuators that generally take advantage of temperature coefficient of expansion (TCE) mismatches in a sandwich structure of two or more materials. An example of bimorph is shown in Figure 1.7.

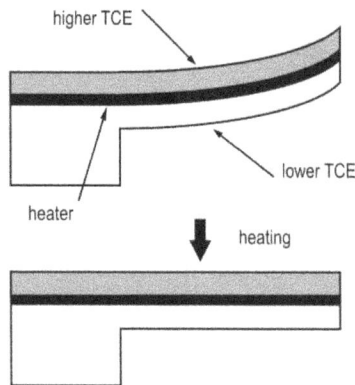

Figure 1.7. Polyimide bimorph. Two kinds of polyimides are used, having different TCE's. A metallic resistive heater exists between the two polyimides.

In this structure, two polyimides with different TCE's are used with an interleaved metal resistive heater. The unheated cantilever is bent upward due to built in stresses, and flattens with application of heat. A variation of a bimorph is a double-armed, doped polysilicon cantilever shown in Figure 1.8. The thinner arm (hot arm) carries a higher current density when electrically biased and heats up more than the thicker arm (cold arm). Hence the difference in thermal expansion leads to actuation.

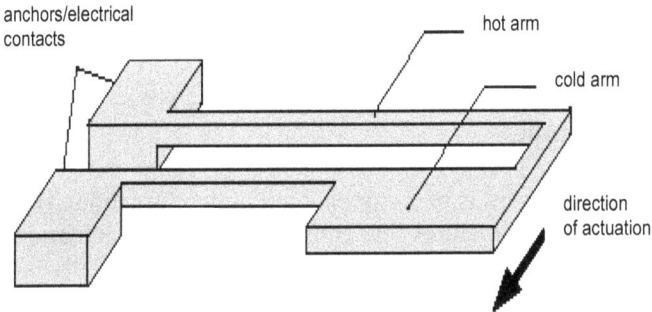

Figure 1.8. Thermally activated "unimorph". Current flows between electrical contacts, giving rise to differential heating in each cantilever arm. Differential heating causes differential thermal expansion and actuation.

1.2.4. Capacitance

Capacitive and electrostatic transduction methods are quite similar, and are distinguished here by the desired function. Capacitive transduction is generally used for sensing, whereas electrostatic transduction is used for actuation. Capacitive micro-machined sensors include pressure sensors, microphones, hydrophones, and accelerometers. They are based on parallel plate capacitors, usually where only one plate is fixed and the other is moving. The capacitance, C, of a parallel plate capacitor is given by

$$C = \frac{\varepsilon A}{d}$$

where ε, A, and d are the permittivity of the gap, the area of the plates, and the separation gap of the plates, respectively. As the gap changes, so does the capacitance. In general, capacitive sensors are more sensitive to mechanical measurands than their piezoresistive counterparts. Also, capacitive sensors generally have less temperature sensitivity. However, micro-machined capacitive sensors typically have capacitances of only a few picofarads, making them susceptible to signal loss through parasitic capacitances. This problem can be mitigated by increasing the area of the sensor, but increases the die size and ultimate cost of the sensor. For these reasons, capacitive sensors have historically been passed over in favor of piezoresistive sensors. However, improvements in analog circuits and their monolithic integration with capacitive sensors have overcome many of the problems and have made capacitive sensors an attractive technology.

1.2.5. Electrostatics

Electrostatic actuation is based upon Coulombic attraction. There are at least four kinds of structures that employ electrostatic actuation: plates, which are pulled to the substrate, rotary electrostatic motors, comb drives, and parallelogram actuators. Plates, shown in Figure 1.9, can be singly clamped beams (cantilevers), doubly clamped beams (bridges), or edge clamped plates (diaphragms). They are distinguished from the other structures because their motion is out of the plane of the substrate, whereas all other structures above are in plane.

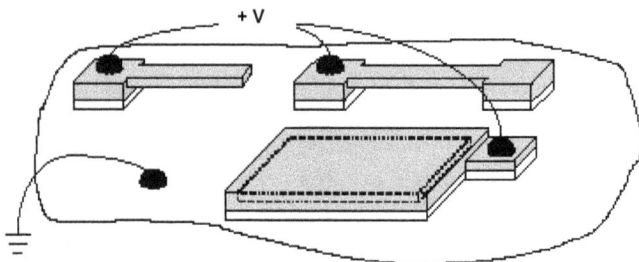

Figure 1.9. Plate electrostatic actuators: (top left) cantilever, (top right) bridge, (bottom) diaphragm.

1.3. Micro-machined Pressure Sensors

This section is a review of previously reported pressure sensors, starting with traditional "macro-machined" and then dealing with micro-machined pressure sensors. Most micro-machined pressure sensors are based on diaphragms. However, several transduction techniques have been used for micro-machined pressure sensors, including piezoresistance, capacitance, optics, and resonance. Since piezoresistance and capacitive sensors are the most common in the literature, the emphasis will be on them. Acoustic sensors, which are often similar to pressure sensors, are also presented.

1.3.1. Macro- and Micro-scale Devices

Here we describe the details of micro-machined pressure sensors; it is instructive to examine some of the macroscopic devices that have been reported. Some of these devices were reported in reference [57], published in 1971, and are shown in Figure 1.10.

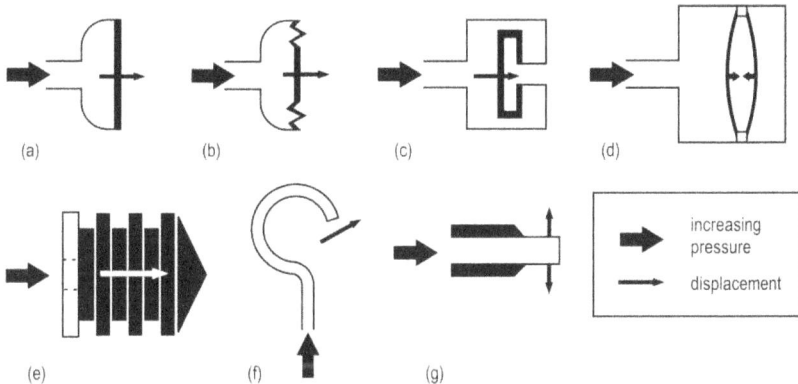

Figure 1.10. Macroscopic pressure sensors: (a) simple diaphragm, (b) corrugated diaphragm, (c) capsule, (d) capacitive sensor, (e) bellows, (f) bourdon tube, and (g) straight tube.

Many of these devices were based on diaphragms (a, b, d). Other devices sought to improve the amount of deflection of a simple diaphragm such as the capsule (c) and bellows (e). Strain gauges were commonly used on diaphragm-based devices. Some diaphragm sensors, however, had elaborate systems of levers, which were linked to electric switches or potentiometer windings. Some diaphragm sensors, instead of having strain gauges mounted directly on the diaphragm itself, have a piston which was driven into a mounted strain gauge by the motion of diaphragm, Finally Bourdon tubes and straight walled tubes, deflected or expanded in the presence of increased pressure (f, g).

Vacuum pressure sensors typically use different transduction mechanism than their greater-than-atmospheric counterparts. A Pirani gauge measures the thermal conductivity of the ambient gas, which is directly proportional to pressure in the 1-2000 mTorr range [57]. A heated resistor is used for this measurement. Ionization gauges operate at pressures from 1mTorr down to $2 \cdot 10^8$ Torr [57]. In these gauges, electrons are emitted from a cathode and accelerated towards an anode plate. Positive ions are created by electron gas collisions. These ions are attracted to a third plate. The current on this plate is proportional to the absolute pressure of the gas.

Many micro-machined pressure sensors are miniaturized versions of their macroscopic counterparts. A micro-machined Pirani gauge has been reported for measuring vacuum [58]. Most sensors for greater-than-atmospheric pressure share the common characteristic of deformable diaphragms. In diaphragm-based sensors, pressure is determined by the deflection of the diaphragm due to applied pressure. Figure 1.11 illustrates a schematic cross-section of a typical pressure sensor diaphragm.

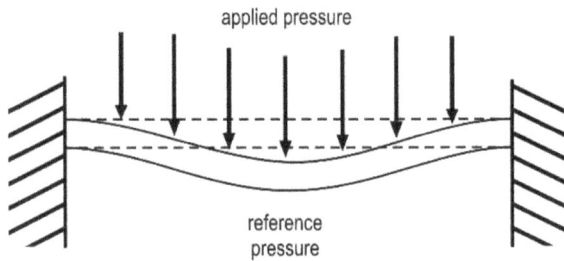

Figure 1.11. Schematic cross section of typical pressure sensor diaphragm.

The reference pressure can be a sealed chamber or a pressure port so that absolute or gauge pressures are measured, respectively. The shape of the diaphragm as viewed from the top is arbitrary, but generally takes the form of a square or circle. These shapes behave similarly to an applied stress. For the case of a clamped circular plate with small deflections (i.e. less than half of the diaphragm thickness) the form of the deflection is

$$w(r) = \frac{P(a^2 - r^2)^2}{64D}$$,

36

where w, r, a, and P are the deflection, radial distance from the center of the diaphragm, diaphragm radius, and applied pressure, respectively. D is the flexural rigidity, given by

$$D = \frac{Eh^3}{12(1-v^2)},$$

where E, h, and v are Young's modulus, thickness and Poisson's ratio, respectively, of the diaphragm. From the above equations, both the shape and the amount of the deflection can be determined. Moreover, it is readily apparent that the amount of deflection is directly proportional to the applied pressure. For the case of a diaphragm with the large built in stress or large deflections this direct proportionality is no longer true. In general, it is desirable to use a deflection measurement scheme that is linear with pressure, since such systems are simple to calibrate and measure.

1.3.2. Piezoresistive Pressure Sensors

Following the invention of the bipolar transistor in 1947, a great deal of effort was put into characterizing the properties of single crystal semiconductors [59]. In 1954, Smith reported the piezoresistive effect of silicon and germanium [60], which enabled semiconductor-based sensors. These types of sensors are sensitive to the induced strain of the diaphragms, which for a clamped circular plate are given by

$$\varepsilon_r = \frac{Pa^2h}{32D}\left(3\left(\frac{r}{a}\right)^2 - 1\right)$$

$$\varepsilon_\theta = \frac{Pa^2h}{32D}\left(\left(\frac{r}{a}\right)^2 - 1\right),$$

where, ε_r, ε_θ and h are the radial strain, circumferential strain, and diaphragm thickness, respectively. From these equations it is seen that the strain varies linearly with applied pressure. Since the piezoresistive coefficients are linear with strain, the net resistance change is linear with pressure.

The evolution of piezoresistive pressure sensor technology is illustrated in Figure 1.12, starting with metal diaphragm sensors with bonded silicon gauges [Figure 1.12 (a)]. The strain gauges were bonded by epoxies, phenolics or eutectics. The first designs had low yield and poor stability due to such things as thermal mismatch with the metal/epoxy/silicon interface.

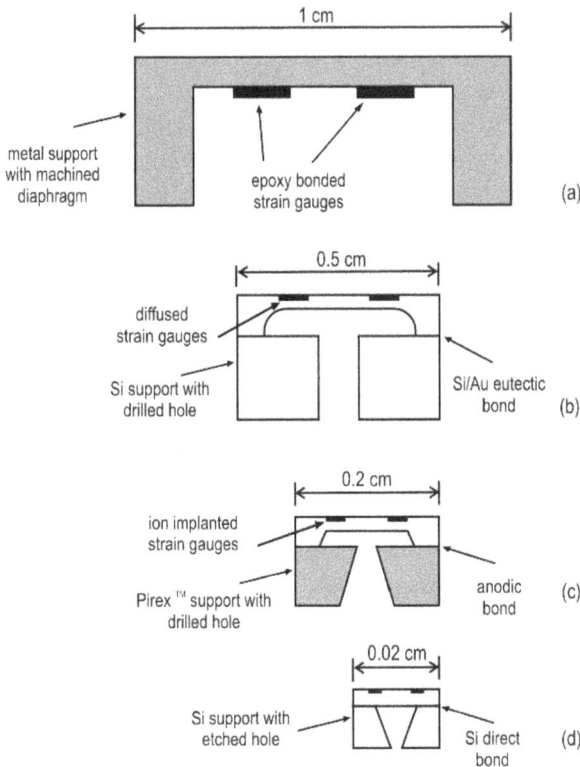

Figure 1.12. Evolution of diaphragm pressure sensors. Adapted from reference [61].

Metal diaphragms were quickly superseded by single crystal diaphragm with diffused piezoresistors [Figure 1.12 (b)]. These types of sensors had many advantages related to the properties of silicon and the availability of high quality silicon substrates. Hysteresis and creep, which were associated with the metal diaphragm and plastic deformation, were eliminated. At low temperatures (< 500 °C), silicon is perfectly elastic and will not plastically deform, but instead will fracture in a brittle manner. Silicon obeys Hooke's law (elastic behavior) up to 1 % strain, a tenfold increase over common alloys. Also, the ultimate tensile strength of silicon can be three times higher than stainless steel wire. As a piezoresistive material, silicon has gauge factors that are over an order of magnitude higher than metal alloys.

By the late 1960's and early 1970's, three key technologies were being developed: anisotropic chemical etching of silicon, ion implantation, and anodic bonding. Ion implantation was used to place strain gauges in single crystal silicon diaphragms. Ion implantation is generally better than diffusion for doping because both the doping concentration and doping uniformity are more tightly controlled. Anisotropic etching improved the diaphragm fabrication process in a number of ways:

1) Diaphragm sizes and locations were now well controlled by IC photolithography techniques;

2) Strain gauge placements were improved;

3) Anisotropic etching was well suited to batch fabrication;

4) Overall size was decreased further.

Anodic bonding was used to bond the finished silicon diaphragm wafers to Pyrex^{TM} glass supports. Several types of glass formulations were used, and were designed to reduce thermal mismatch to silicon. Anisotropic etching and anodic bonding were batch techniques, and hence hundreds (or more) of pressure sensors could be manufactured simultaneously on a single wafer. This amounted to a significant cost reduction. A representative sensor from this is shown in Figure 1.12 (c).

The 1980's to the present has been called the micromachining period, since diaphragm dimensions are shrinking to hundreds of microns or less and minimum feature sizes are shrinking to microns (Figure 1.12(d)). Also, anisotropic etching and bonding technologies are being improved. In 1985, the direct bonding method was first reported. This method was first used for making silicon-on-insulator (SOI) material, but was quickly applied to micro-machined devices. Also, surface micro-machined devices have been reported, which have silicon nitride or polysilicon diaphragms. These sensors decrease required die size and may simplify integration with electronics, but at the cost of reduced sensitivity and reproducibility of mechanical properties.

1.3.3. Capacitive Pressure Sensors

Capacitive pressure sensors are based upon parallel plate capacitors. A typical bulk micro-machined capacitive pressure sensor is shown in Figure 1.13.

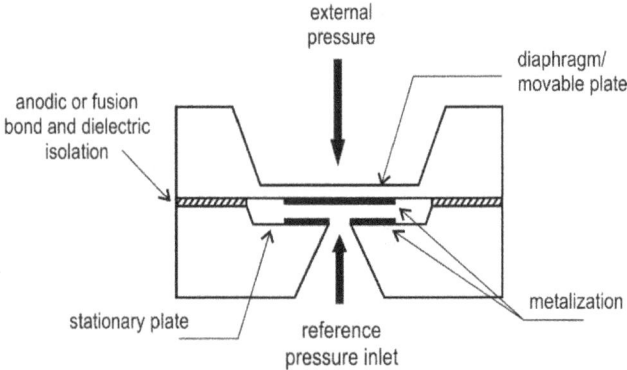

Figure 1.13. Cross section schematic of bulk micro-machined capacitive pressure sensor. Adapted from reference.

The capacitance, C, of a parallel plate capacitor is given by

$$C = \frac{\varepsilon A}{d}$$

where ε, A, and d are the permittivity of the gap, the area of the plates, and the separation of the plates, respectively. For a moving, circular diaphragm sensor, the capacitance becomes

$$C = \iint \frac{\varepsilon}{d - w(r,\theta)} r\, dr\, d\theta \quad,$$

where w is the deflection of the diaphragm. The capacitance with respect to applied pressure is generally nonlinear due to the nonlinear deflected shape of the diaphragm. A possibility for increased linearity is to operate a capacitive sensor in contact mode (Figure 1.14).

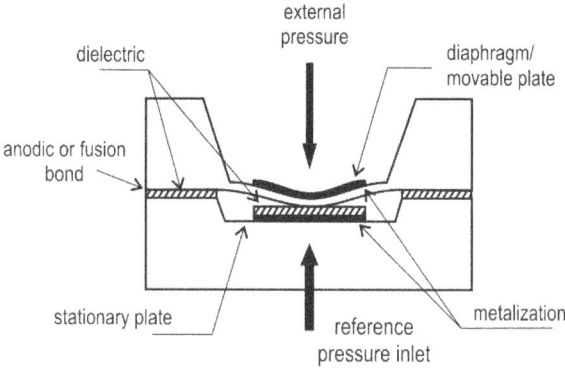

Figure 1.14. Cross section schematic of bulk micro-machined, contact-mode pressure sensor.

In contact mode, the capacitance is directly proportional to the contact area which in turn exhibits good linearity with respect to applied pressure. This holds true over a wide range of pressures. However, this linearity comes at the expense of decreased sensitivity. However, in contrast to a piezoresistive pressure sensor, a capacitive pressure sensor will have a monotonic increase of capacitance with applied pressure, even in contact mode.

One method for achieving a linear response is to use bossed diaphragms. Figure 1.15 illustrates this concept. On the left is a cut-away view of a uniform thickness diaphragm and its corresponding cross section deflected mode shape. A non-uniform, bossed diaphragm is on the right.

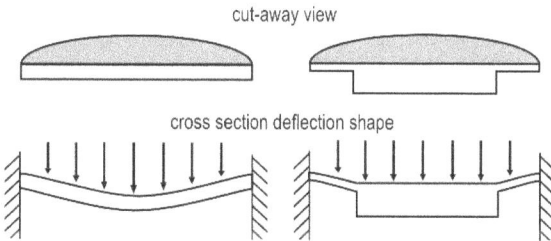

Figure 1.15. Comparison of deflection shapes for uniform thickness (left) and bossed (right) diaphragms.

The thicker center portion is much stiffer than the thinner tether portion on the outside. The center boss contributes most of the capacitance of the structure and its shape does not distort appreciably under applied load. Hence the capacitance-pressure characteristics will be more linear. The principal advantages of capacitive pressure sensors over piezoresistive pressure sensors are increased pressure sensitivity and decreased temperature sensitivity.

1.3.4. Piezoelectric Pressure Sensors

Piezoelectric pressure sensors have not been prevalent in the literature, since piezoelectric materials are not suitable for static or slowly varying measurements. However, various acoustic sensors based upon piezoelectric materials, have been reported, which are closely related to pressure sensors. Pressure sensors have been reported which are excite and measure the resonant frequency of a diaphragm as a function of pressure. Although the resonant frequency varies as the square root of the pressure, these devices display good linearity for low pressures. Also, polysilicon diaphragms with ZnO deposited on top have been made which exhibit excellent linearity over a wide range of pressures.

1.3.5. Optical Sensors

Many diaphragm based optical sensors have been reported which measure pressure induced deflections by Mach Zender interferometery and Fabry Perot interferometery. The deflection derived from these devices varies linearly with pressure. Optical sensors can be quite accurate, but often suffer from temperature sensitivity problems. Furthermore, aligning the optics and calibrating the sensors can be challenging.

References

[1]. S. D. Senturia, Feynman Revisited, in *Proceedings of the IEEE Micro – Electro – Mechanical –Systems Workshop,* Osio, Japan, 25 –28 January, 1994, pp. 309 – 312.
[2]. J. Akhtar, B. B. Dixit, B. D. Pant, and V. P. Deshwal, Polysilicon piezoresistive pressure sensors based on MEMS technology, *IETE Journal of Research,* Vol. 49, No. 6, November-December, 2003, pp. 365-377.

[3]. K. H. L. Chau, S. R. Lewis, R. Zhao, R. T. Howe, S. F. Bart, and R. Marchelli, An integrated force balanced capacitive accelerometer for low – g applications, in *Technical Digest of 8th International Conference on Solid State Sensors and Actuators*, 1995, pp. 593-596.

[4]. A. V. Chavan and K. D. Wise, A batch process vacuum sealed capacitive pressure sensor, in *Technical Digest of International Conference on Solid State Sensors and Actuators*, 1997, pp. 1449-1452.

[5]. J. Chen and K. Wise, A high resolution silicon monolithic nozzle array inkjet printing, in *Technical Digest of 8th International Conference on Solid State Sensors and Actuators*, 1995, pp. 321-324.

[6]. P. Gravesen, J. Branebjerg and O. S. Jensen, Micro – fluidics: A Review, *Journal of Micromechanics and Microengineering*, 1992, pp. 168-182.

[7]. K. Peterson, W. McMillan, G. Kovac, A. Northruo, L. Christel and F. Pourahmadi, The promise of miniaturized clinical diagnosis systems, *IVD Technology*, July, 1998, pp. 154-168.

[8]. S. Lee, E. Mottamedi and M. C. Wu, Surface micro – machined free space fiber optic switches with integrated micro actuators for optical fiber communication systems, in *Technical Digest of International Conference on Solid State Sensors and Actuators*, 1997, pp. 85-88.

[9]. G. M. Rebeiz and J. B. Muldavin, RF MEMS switches circuits, *IEEE Microwave Magazine*, Vol. 2, No. 4, December, 2001, pp. 59-71.

[10]. Marc J. Madou, Fundamental of micro fabrication, 2nd ed., *CRC Press*, NY, 2001.

[11]. C. S. Smith, Piezoelectric effect in germanium and silicon, *Physical Review*, 94, 1, 1954, pp. 42-49.

[12]. J. Bryzek, Characterization of MEMS industry in silicon valley and its impact on sensor technology, in *Proceedings of the Sensor Expo' 96*, Philadelphia, PA, October, 1996, pp. 1-13.

[13]. J. Bryzek, MEMS: A closer look, Part 2: the MEMS industry in silicon valley and its impact on sensor technology, *Sensors*, July, 1996, p. 4.

[14]. H. C. Nathanson and Wickstrom, A resonant gate silicon surface transistor with high-Q band pass properties, *Applied Physics Letters*, 7, 1965, p. 84.

[15]. K. E. Petersen, Silicon as a mechanical material, *Proceedings of the IEEE*, 70, 5, May, 1982, pp. 420-457.

[16]. H. Robbins and B. Schwartz, Chemical etching of silicon, I. The system HF, HNO_3, and H_2O, *Journal of the Electrochemical Society*, 106, 1959, p. 505.

[17]. H. Robbins and B. Schwartz, Chemical etching of silicon, II. The system HF, HNO_3, and H_2O, and $HC_2H_3O_2$, *Journal of the Electrochemical Society*, 107, 1960, p. 108.

[18]. B. Schwartz and H. Robbins, Chemical etching of silicon, III. A temperature study in the acid system, *Journal of the Electrochemical Society*, 108, 1961, p. 365.

[19]. B. Schwartz and H. Robbins, Chemical etching of silicon, IV. Etching technology, *Journal of the Electrochemical Society*, 123, 1976, p. 1903.

[20]. D. L. Kendall, On etching very narrow grooves in silicon, *Applied Physics Letters*, 26, 1975, p. 195.

[21]. D. L. Kendall, Vertical etching of silicon at very high aspect ratios, *Annual Review of Material Science*, 9, 1979, pp. 373-403.

[22]. D. L. Kendall, A new theory for the anisotropic etching of silicon and some underdeveloped chemical micromachining concepts, *Journal of Vacuum Science Technology A*, 8, 4, 1990, pp. 3598-3604.

[23]. K. Bean, Anisotropic etching of silicon, *IEEE Transactions on Electron Devices*, ED-25, 1978, pp. 1185-1193.

[24]. E. Bassous, Fabrications of novel three-dimensional microstructures by the anisotropic etching of (100) and (110) silicon, *IEEE Transactions on Electron Devices*, ED-25, 1978, p. 1178.

[25]. A. C. M. Gieles and G. H. J. Somers, Miniature pressure transducers with a silicon diaphragm, *Philip Technical Review*, 33, 1, 1973, pp. 14-20.

[26]. J. Bryzek, K. Petersen, J. R. Mallon, L. Christel, and F. Pourahmadi, *Silicon Sensors and Microstructures*, Novasensor, Fremont, CA, 1990.

[27]. Sonic Mill, Albuquerque, NM.

[28]. T. M. Bloomstein, Laser deposition and etching of three-dimensional microstructures, Digest of Technical Papers, in *Proceedings of the International Conference on Solid-State Sensors and Actuators, Transducers*, 1991, pp. 507-511.

[29]. M. Alavi, S. Buttgenbach, A. Schumacher, H.–J Wagner, Laser machining of silicon for fabrication of new microstructures, Digest of Technical Papers, *International Conference on Solid-State Sensors and Actuators, Transducers*, 1991, pp. 512-515.

[30]. T. Masaki, K. Kawata, and T. Masuzawa, Micro electro-discharge machining and its applications, in *Proceedings of IEEE Micro Electro Mechanical Systems*, 1990, pp. 21-26.

[31]. Y. Backlund, and L. Rosengren, New shapes in (100) Si using KOH and EDP etches, *Journal of Micromechanics and Microengineering*, 2, 2, June, 1992, pp. 75-79.

[32]. M. J. Declercq, L. Gerzberg, and J. D. Meindl, Optimization of the hydrazine water solution for anisotropic etching of silicon in integrated circuit technology, *Journal of Electrochemical Society*, 122, 1975, pp. 545-552.

[33]. F. A. Chambers, and L. S. Wilkiel, Cesium hydroxide etching of (100) silicon, *Journal of Micromechanics and Microengineering*, 3, 1993, pp. 1-3.

[34]. U. Schnakenberg, W. Benecke, and B. Lochel, NH_4OH based etchants for silicon micromachining, *Sensors and Actuators A*, A21-A23, 1989, pp. 1031-1035.

[35]. W.-S. Choi and J. G. Smits, A method to etch undoped silicon cantilever beams, *Journal of Micromechanical Systems*, 2, 2, June, 1993, pp. 82-86.

[36]. O. Tabata, R. Asahi, H. Funabashi, and S. Sugiyama, Anisotropic etching of silicon in $(CH_3)_4NOH$ solutions, *Digest of Technical Papers,*

International Conference on Solid - State Sensors and Actuators, Transducers, 1991, pp. 811-814.

[37]. U. Schnakenberg, W. Benecke, and P. Lange, TMAHW etchants for silicon Micromachining, Digest of Technical Papers, in *Proceedings of the International Conference on Solid - State Sensors and Actuators, Transducers*, 1991, pp. 815-817.

[38]. M Aslam, B. E. Artz, S. L. Kaberline, T. J. Prater, A comparison of cleaning procedures for removing potassium from wafers exposed to KOH, *IEEE Transactions on Electron Devices*, 40, 2, February, 1993, pp. 292-5.

[39]. P. J. French, Development of surface micromachining techniques compatible with on-chip electronics, *Journal of Micromechanics and Microengineering*, 6, 1996, 197-211.

[40]. E. W. Becker, W. Ehrfeld, P. Hagmann, A. Maner and D. Münchmeyer, Fabrication of microstructures with high aspect ratios and great structural heights by synchrotron radiation, lithography, galvanoformung and plastic molding (LIGA Process), *Microelectronic Engineering*, 4, 1986, pp. 33-56.

[41]. H. Guckel, K. J. Skrobis, T. R. Christenson, J. Klein, S. Han, B. Choi and E. J. Lovell, Fabrication of assembled micromachined components via deep x-ray lithography, in *Proceedings of IEEE MEMS*, 91, 1991, pp. 74-79.

[42]. C. H. Mastrangelo and W. C. Tang, Semiconductor sensor technologies, in Semiconductor Sensors, Ed. S. M. Sze, *John Wiley & Sons*, 1994, pp. 17-84.

[43]. H. Guckel, K. J. Skrobis, T. R. Christenson, J. Klein, S. Han, B. Choi, E. G. Lovell, T. W. Chapman, On the application of deep x-ray lithography with sacrificial layers to sensor and actuator construction (the magnetic micromotor with power takeoffs), *Digest of Technical Papers, International Conference on Solid-State Sensors and Actuators, Transducers*, 1991.

[44]. W. Ehrfeld, P. Bley, F. Götz, J. Mohr, D. Münchmeyer, W. Schelb, H. J. Baving, and D. Beets, Progress in deep-etch synchrotron radiation lithography, *Journal of Vacuum Science and Technology*, B, 6, 1, Jan/Feb, 1988, pp. 178-182.

[45]. T. R. Christenson and H. Guckel, Deep x-ray lithography for micromechanics, in *Proceedings of the SPIE Symposium on Micromachining and Microfabrication Process Technology*, Austin, TX, 2639, Oct., 1995, pp. 136-145.

[46]. R. Ruprecht, W. Bacher, J. H. Haubelt, V. Piotter, Injection molding of LIGA and LIGA –similar microstructures using filled and unfilled thermoplastics, in *Proceedings of the SPIE, Micromachining and Microfabrication Process Technology*, 2639, Oct., 1995, pp. 146-157.

[47]. B. Loechel, A. Maciossek, Surface microcomponents fabricated by UV depth lithography and electroplating, in *Proceedings of the SPIE*,

Micromachining and Microfabrication Process Technology, 2639, Oct., 1995, pp. 174-184.

[48]. H. Miyajima and M. Mehregany, High aspect ratio photolithography for MEMS applications, *Journal of Microelectromechanical Systems*, 4, 4, Dec., 1995, pp. 220-229.

[49]. K. A. Shaw, Z. L. Zhang, and N. C. MacDonald, SCREAM I: A single-crystal silicon process for microelectromechanical structures, in *Proceedings of IEEE Microelectromechanical Systems*, Feb., 1993, pp. 155-160.

[50]. C. Keller and M. Ferrari, Milli-scale polysilicon structures, *Technical Digest: Solid State Sensor and Actuator Workshop*, Hilton Head, 1994, pp. 132-137.

[51]. J. G. Fleming and C. C. Barron, Novel silicon fabrication process for high aspect ratio micromachined parts, in *Proceedings of the SPIE, Micromachining and Microfabrication Process Technology*, 2639, Oct., 1995, pp. 185-190.

[52]. D. Sander, R. Hoffman, V. Relling, and J. Müller, Fabrication of metallic microstructures by electroplating using deep-etched silicon molds, *Journal of Microelectromechanical Systems*, 4, 2, June, 1995, pp. 81-86.

[53]. B. Kloeck and N. F. de Rooij, Mechanical Sensors, in Semiconductor Sensors, Ed. S. M. Sze, *John Willey & Sons*, 1994, pp. 153-199.

[54]. C. S. Smith, Piezoresistance effect in germanium and silicon, *Physical Review*, 94, 1, 1954, pp. 42-49.

[55]. M. Akbar and M. A. Shanblatt, A fully integrated temperature compensation technique for piezoresistive pressure sensor, *IEEE Transactions on Electronic Devices*, ED-42, 3, June, 1993, pp. 771-775.

[56]. M.-H. Bao, L.-Z. Yu and Y. Wang, Micromachined beam-diaphragm structure improves performance of pressure transducer, *Sensors and Actuators A*, A21-A23, 1990, pp. 137-141.

[57]. F. J. Oliver, Practical Instrumentation Transducers, *Hayden Book Company, Inc*, New York, 1971, pp. 146-172.

[58]. C. H. Mastrangelo, Thermal Applications of Microbridges, Ph. D. Thesis, *University of California at Berkeley*, 1991.

[59]. S. M Sze, Classification and Terminology of Sensors, in Semiconductor Sensors, Ed. S. M. Sze, *John Wiley & Sons*, 1994, pp. 1-15.

[60]. C. S. Smith, Piezoresistance effect in germanium and silicon, *Physical Review*, 94, 1, 1954, pp. 42-49.

[61]. J. Bryzek, K. Petersen, J. R. Mellon, L. Christel, and F. Pourahmadi, Silicon sensors and microstructures, *Novasensor*, Fremont, CA, 1990.

Chapter 2

Micro-fabrication Technologies

2.1. Micro-fabrication Techniques

Micro-fabrication, as practiced in the microelectronics and MEMS fields, is based on planer technologies: constructing the electronics devices and MEMS components on substrates that are in the form of initially flat wafers. Because the microelectronics industry has made huge investments to develop wafer-based process technologies, there is a correspondingly huge advantage for MEMS designers to exploit these same process steps, or variants based on those steps. Here, some of the micro-fabrication techniques are presented.

2.2. Silicon Wafers

Planar substrates of choice include single – crystal silicon, single – crystal quartz, glass, and fused (amorphous) quartz. All are available in wafer form in sizes that are compatible with standard microelectronics processing equipments. As the microelectronics industry moves toward larger and larger wafer sizes [200 mm (8″) diameter wafers are now standard], there is a pressure on MEMS fabricators to shift to increasing wafer sizes to maintain compatibility with production equipment. However, MEMS fabrication obeys different economics than standard microelectronics.

Single – crystal silicon wafers are classified by the orientation of the surface relative to the crystalline axes. The orientation of a silicon crystal is an important parameter in the device fabrication sequence. One method used to describe the orientation of crystals is through the use of Miller indices. The nomenclature is based on the miller indices [1-3]. Another approach i.e. industrial approach to identify the crystal orientation of wafers is shown in Figure 2.1.

Wafer Identification

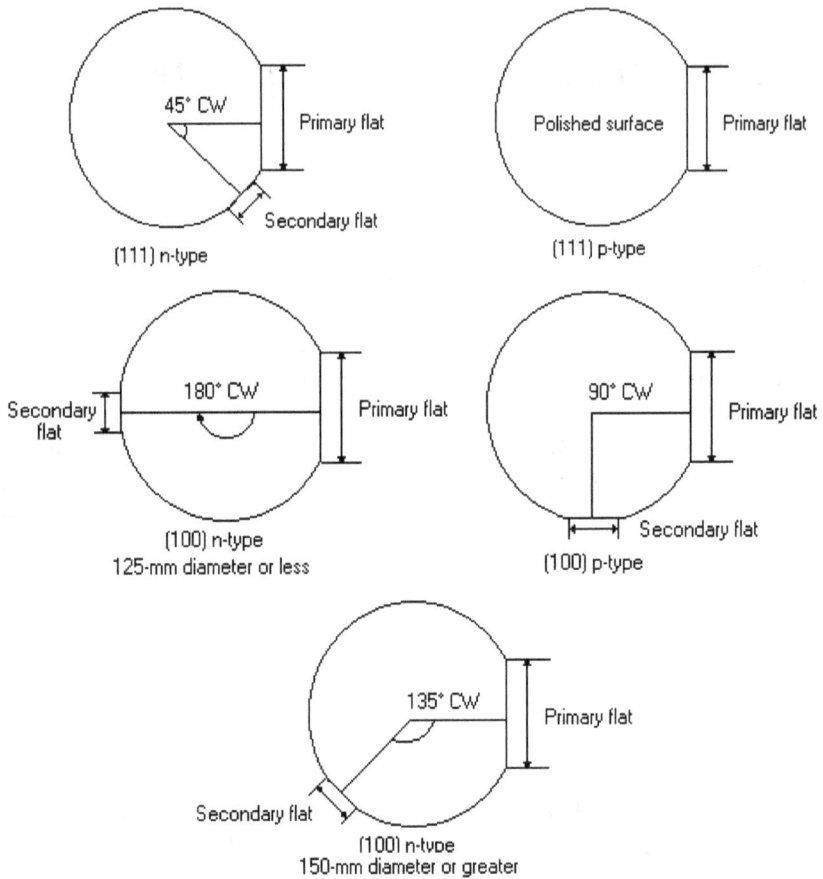

Figure 2.1. The industrial way of identifying wafers.

2.3. Wafers Cleaning

One of the first considerations that must be made in wafer processing is on how wafers can be cleaned prior to any process step. Attempts are made to keep wafers clean at all times, but more care must be taken prior to high-temperature processing steps such as diffusion, epitaxial growth, or chemical vapor deposition. The two types of contamination that have been found to cause the largest problems in semiconductors are ions that are mobile in silicon dioxide, e.g. sodium and elements that diffuse in silicon and precipitate out somewhere in the interior, e.g. gold and some other metals.

Sodium interfaces with the normal operation of semiconductor devices by rapidly drifting through silicon dioxide toward regions with a negative bias. It then gives rise to changes in device characteristics such as excessive leakage. Sodium can be kept out of a fabrication line by specifying low-sodium chemicals and by rigorously enforcing proper handling techniques. Sodium is a chemical present in the human body, and careless procedures will result in unwanted sodium contamination of wafers or wafer-handling equipment.

Certain elements are soluble in silicon dioxide at elevated temperatures, but precipitate into non-lattice locations when the wafer temperature is lowered. These elements interfere with the normal flow of holes and electrons in the silicon crystal when the device is in operation. Once a quantity of any of these elements has contaminated a wafer, it is impossible to completely remove it. However, proper cleaning prior to a high temperature operation minimizes its effect.

The first step in dealing with wafers is to thoroughly degrease them.

2.3.1. Degreasing

1) Dip the wafer in 1, 1, 1- trichloroethane (TCE), boil for 10 minutes and place in Ultrasonic radiation for 5 minutes.

2) Dip the wafer in Acetone, boil for 10 minutes and place in Ultrasonic radiation for 5 minutes.

3) Dip the wafer in Methanol, boil for 10 minutes and place in Ultrasonic radiation for 5 minutes.

4) Rinse the wafer in De-ionized water, DI (optional).

5) Finally dry the wafer.

This cleaning procedure guarantees that any greases or waxes that might be insoluble in subsequent cleaning steps are removed. Wafers are then sent through a series of solutions designed to remove any trace of metals or other potentially harmful materials. There are different methods for wafer cleaning that are in use. Except for environmental concerns, wet treatment still outranks other cleaning procedures.

2.3.2. RCA 1

This solution contains DI water: H_2O_2 : NH_4OH in the ratio 5:1:1. Wafers are immersed in this solution and boiled for about 10 minutes followed by thorough rinse in DI water. This procedure removes organic dirt (resist).

2.3.3. RCA 2

This solution contains DI water: H_2O_2 : HCl in the ratio 6:1:1. Wafers are immersed in this solution and boiled for about 10 minutes followed by thorough rinse in DI water. This procedure removes metal ions.

Both cleaning processes leave a thin oxide on the wafers. This oxide must be stripped of by dipping the wafers in a solution of DI water: HF in the ratio 50:1 for a very short time. One important thing to keep in mind is that HF reacts with the glass so for this solution Teflon beaker must be used.

Another cleaning which comes under wet treatment is the Piranha treatment. This one serves the above two purposes, i.e. that of RCA 1 and RCA 2.

2.3.4. Piranha Treatment

This solution contains H_2SO_4 : H_2O_2: in the ratio 7:1. H_2O_2 must be poured in the beaker first. Wafers are immersed in this solution for about 10 minutes followed by dipping in a solution of DI water: HF in the ratio 50:1 for a very short time. Then the wafers are rinsed in DI water and then dried. But RCA 1 and RCA 2 cleaning is more preferred over piranha because of the high risk involved in piranha process like the solution is highly volatile and also this process can cause damage to some very delicate devices. Piranha is used where RCA does not work and cleaning is required quickly. Figure 2.2 shows etching tanks used to perform Piranha, Hydrofluoric acid or RCA clean on 4-inch wafer batches at LAAS technological facility in Toulouse, France.

Figure 2.2. Etching tank for wafer cleaning.

2.4 Thermal Oxidation

Silicon form several oxides, the most important of which is silicon dioxide (SiO_2). High quality, thick SiO_2 films may be grown on a silicon substrate by thermal oxidation. Since the silicon surface has a high affinity to an oxidizing ambient, thermal oxidation may obtain in oxidation furnace in a temperature range of 600 to 1250 °C.

When dry or wet oxygen is introduced into the chamber at a constant rate, the oxidation is obtained following two reactions:

$$Si + O_2 \longrightarrow SiO_2 \text{ (dry process)}$$

$$Si + 2H_2O \longrightarrow SiO_2 + 2H_2 \text{ (wet process)}$$

In this process, the silicon at the surface of the wafer combines with oxygen gas to form SiO_2. The Si/ SiO_2 interface is moved into silicon, so the silicon slowly recedes from its original location [4].

Figure 2.3 and 2.4 shows a typical oxidation apparatus. The wafers are placed in a fused silica rack called a wafer boat. The wafer boat slowly inserted into a fused silica tube wrapped in an electrical mental. The temperature of the wafers gradually rises as the wafer boat moves into the middle of the heating zone. Oxygen gas blowing through the tube passes over the surface of the each wafer. At elevated temperatures, oxygen molecules can actually diffused through the oxide layer to reach the underlying silicon. There oxygen and silicon reacts and layer of oxide gradually grows thicker. The rate of oxygen diffusion slows as the oxide film thickness, so the growth rate decreases with time.

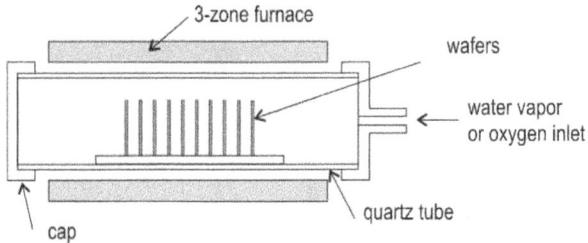

Figure 2.3. Tube furnace for thermal oxidation.

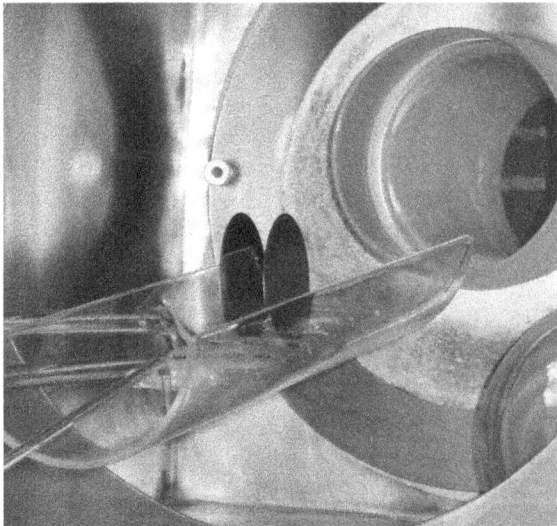

Figure 2.4. Oxidation furnace with loading of wafers holding quartz boat at CEERI, Pilani (India).

High temperature greatly accelerates oxide growth. Crystal orientation also affects oxidation rates, with (111) silicon oxidizing significantly faster than (100) silicon. Once the oxide layer has reached the desired thickness, the wafers are slowly withdrawn from the furnace.

Dry oxide grows very slowly, but it is of particularly high quality because relatively few defects exist at the oxide – silicon interface. These defects, or surface states, interface with the proper operation of semiconductor devices, particularly MOS transistors. The density of surface states is measured by a parameter called the surface state charge. Dry oxide films that are thermally grown on (100) silicon have especially low surface state charges and thus make ideal dielectrics for MOS transistors. Wet oxides are formed in the same way as dry oxides, but steam is injected into the furnace tube to accelerate the oxidation. Water vapor moves rapidly through oxide films, but hydrogen atoms liberated by the decomposition of the water molecules produce imperfections that may degrade the oxide quality. Wet oxidation is commonly used to grow a thick layer of oxide. Dry oxidations conducted at higher than ambient pressures can also accelerate oxide growth rates.

A comparison between dry and wet oxidation is given below:

Dry Oxidation	Wet Oxidation
$Si + O_2 \rightarrow SiO_2$	$Si + 2H_2O \rightarrow SiO_2 + 2H_2$
Dense oxide formed	Relatively porous oxide formed
Good quality, low diffusion	Lower quality, species diffuse faster
Slow growth rate	Faster growth rate

The thermal oxidation rate is influenced by the orientation of the silicon substrate. The linear oxidation rate for silicon follows the sequence: (1 1 0) > (1 1 1) > (1 0 0), corresponding to increasing activation energy incorporated a term for the bond density in the plane as one the bond orientation.

2.5. CVD Techniques

Chemical Vapor Deposition (CVD) processes have been originated in very large – scale integration (VLSI) microelectronics, and the principle of deposition is based on a surface chemical reaction of one or more reactant gases, where the thermal energy source is solely based on high temperature reactors such as atmospheric pressure CVD (APCVD), LPCVD, or an additional energy source such as electrically excited plasma used by PECVD (plasma enhanced).

In CVD techniques, the substrate is placed inside a reactor to which a number of gases are supplied. The fundamental principle of the process is that a chemical reaction takes place between the source gases. The product of that reaction is a solid material, which condenses on all surfaces inside the reactor. The two most important CVD technologies in MEMS are the Low Pressure CVD (LPCVD) and Plasma Enhanced CVD (PECVD).

2.5.1. LPCVD (Low pressure Chemical Vapor Deposition)

The LPCVD process is carried out in a high temperature furnace where the surface reaction is obtained with a thermal energy source. LPCVD Silicon nitride and silicon oxinitride films are grown at low pressure, below 130 Pa, and high temperatures between 750 and 950 °C [5]. Under these processing conditions, nitrides and oxinitrides are stochiometric and mostly amorphous. The LPCVD is chosen to allow surface catalyzed reaction of an otherwise highly reactive gas mixture. Figure 2.5 illustrates a horizontal "hot wall" LPCVD reactor where substrates are vertically placed on a quartz support.

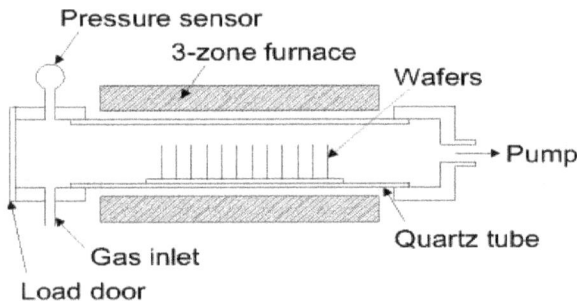

Figure 2.5. Horizontal "hot-wall" LPCVD reactor.

Processing gases are introduced from one extremity of the tube reactor, heated prior to arrival at the substrate by contact with the walls, and then absorbed on the surface when reacting to form the deposited thin film. Usually, two different processing gases, one the silicon-containing precursor and the other the nitrogen-containing source gas, are used in deposition of silicon nitride layers [6]. The usual silicon containing gases are silane and dichlorosilane (SiH_2Cl_2). The usual nitrogen-containing sources are ammonia and pure nitrogen. Silicon oxinitride may be considered as a mixture of Si – O and Si – N bonds, where the refractive index can vary from that of silica (1.46) and those of silicon nitride (~2). Reacting silane or dichlorosilane and several mixtures of gaseous reactants such as ammonia or nitrous oxide can produce LPCVD grown silicon oxinitride [7]. LPCVD silicon dioxide may be deposited by reacting silane or dichlorosilane with nitrous oxide. Successful deposition of silicon nitride films was reported for both $SiH_4 - NH_3$ and $SiH_2Cl_2 - NH_3$ gas mixtures [5, 8].

2.5.2. PECVD (Plasma Enhanced Chemical Vapor Deposition)

In this deposition technique, plasma is used to enhance the breakdown of the gases at low temperature (~300 °C). In the PECVD, radio frequency (RF)-induced plasma activation provides thin film deposition at low operational temperatures because the transfer energy needed for deposition is transferred from plasma to reactant gases. The discharge ionizes the gases, creating radicals that react at the wafer surface. Typically, pure silica, silicon nitride, silicon oxinitride thin films can be fabricated using a conventional parallel-plate PECVD reactor. Substrate can be placed horizontally on a heated susceptor, which acts as one pair of RF electrodes [3]. A mixture of four processing gases, namely, silane, nitrogen, ammonia, and nitrous oxide, are supplied from a flow control system into the region between electrodes generating RF- induced plasma in this region. PECVD pure silica films may be deposited by reacting silane (SiH_4) and nitrous oxide (N_2O) in argon plasma. N_2O produces atomic oxygen that reacts with SiH_4 to form SiO_2 [9]. The PECVD process exhibits a disadvantage based on the presence of large hydrogen content owing to the dissociation of silane in the plasma. Hydrogen is bonded to the silicon as Si – H bonds and the PECVD silicon nitride is in general an amorphous hydrogenated material (a-Si_3N_4: H) [10]. This is usually deposited from a mixture of SiH_4 and NH_3, often diluted with a carrier gas such as He, Ar, or N_2 [9]. It was demonstrated that the deposition of PECVD silicon nitride fabricated from a mixture of 2 % SiH_4 in N_2, where N_2 instead of NH_3 is used as the source of nitrogen, permits the H content of deposited films to be reduced [11]. Even with low H content, PECVD silicon nitride films have a mechanical stress, which may lead to cracking. Heat treatments after deposition could lead to a decrease in cracking resistance in the nitride films [12]. By introducing oxygen, the silicon nitride may be converted to silicon oxinitride. The schematic diagram of PECVD reactor is shown in Figure 2.6.

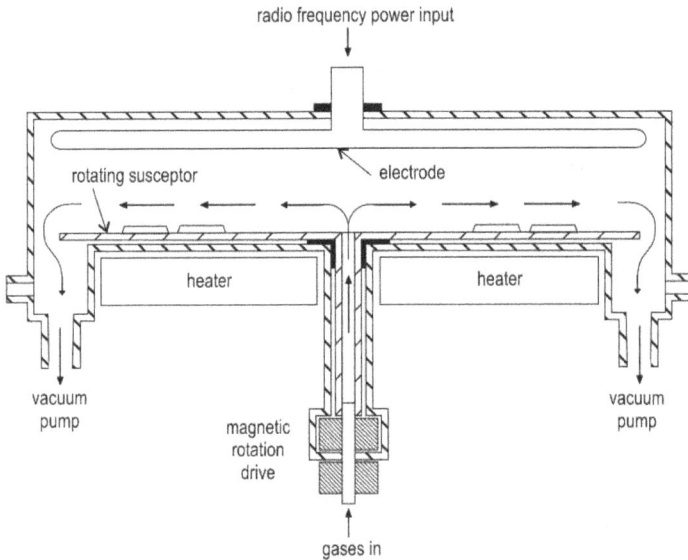

Figure 2.6. Plasma Enhanced CVD system.

A major advantage of LPCVD over PECVD is the higher uniformity and better reproducibility of deposited films. This is partially due to the low content of atomic hydrogen in the deposited thin films. Two main disadvantages of LPCVD are the low deposition rate and relatively high operating temperatures. Two major attributes of PECVD are the relatively high average deposition rate (30-40 nm/min) and the low temperature processing in the range of 300 to 400 °C. Such a range of processing temperatures is compatible with well-established microelectronic processing.

2.5.3. APCVD (Atmospheric Pressure Chemical Vapor Deposition)

APCVD is usually conducted in cold wall reactors, since many of the reactions involved in the deposition of thin films are pyrolytic in character. Resistive heating of the susceptor is often used, since growth temperatures are usually lower than those encountered in vapor-phase epitaxy (VPE). As the name indicates this process is carried out at atmospheric pressure.

2.6. Photolithography

Photolithography is the patterning process that sets the horizontal dimensions on the various parts of the devices and circuits. The goal of the operation is twofold. First, is to create in and on the wafer surface a pattern whose dimensions are as close to the design requirements as possible. This goal is referred to as the resolution of the images on the wafer. The pattern dimensions are referred to as the feature sizes or image sizes of the circuit. The second goal is the correct placement (called alignment) of the circuit pattern on the wafer. The entire circuit pattern must be correctly placed on the wafer surface and the individual parts of the circuit must be in the correct positions relative to each other. The final pattern is created from several photo masks applied to the wafer in a sequential manner.

2.6.1. Wafer Alignment and Wafer Exposure

In the whole processes of MEMS fabrication wafer alignment is the most important step. Since MEMS fabrication deals with device dimensions in microns, a slight misalignment by just a micron will cause great distortions leading to wastage of the device so great care must be taken during wafer alignment. For this purpose, alignment marks are provided on the mask. These alignment marks are the features where the aligner must concentrate while aligning. Dimensions of the alignment marks are usually kept bigger so that they are easily visible thus reducing difficulty in aligning.

After proper alignment, the wafer is exposed to UV rays. This alignment and exposure is done by Mask Aligner (Mask Aligner used in MEMS laboratory, CEERI, Pilani, CSIR, INDIA is *Karl Suss MA56*). In this equipment the exposure time to UV rays can be adjusted as per the requirement.

There are two necessary steps before the alignment and exposure:

1). **Coating of photoresist.** Coating of photoresist is done by the photoresist spinner. The high-speed spin results in uniform spreading of the PR on the wafer. The general spin rate used is 4500 rpm for 30 seconds. With PR-1813, spin rate 4500 rpm for 30 seconds results in uniform PR thickness of about 1.2 micron.

2). **Soft baking.** In the soft baking, the PR coated wafer is kept in an oven at 90 °C for about 30 minutes, as this increase the strength of the PR and its adhesiveness on the wafer.

In photolithography, wavelengths of the light source used for exposure of the resist-coated wafer ranges from deep UV (DUV) i.e., 150 nm to 300 nm, to near UV, i.e., 350 nm to 500 nm. In near UV, one typically uses the g-line (436 nm) or I-line (365 nm) of a mercury lamp. The brightness of shorter wavelength sources is severely reduced compared to longer wavelength sources and the addition of lenses further reduces the efficiency of the exposure system. The radiation induces a chemical reaction in the exposed areas of the photoresist, altering the solubility of the resist in a solvent either directly or indirectly via a sensitizer.

There are two steps that follow alignment and exposure:

1). **Development.** Development transforms the latent resist image formed during exposure into a relief image that will serve as a mask for further additional steps. There are two main technologies for the development – wet development and dry development.

Wet development is simple and easy to implement. Use of solvents in wet development (immersion and spray developers) leads to some swelling of the resist and a loss of adhesion of the resist to the substrate. After developing the wafer needs to be dried using N_2 gun.

Dry development is costly and more complex, but this could overcome the problems of wet development as it is based on a vapor phase process or plasma. This one is preferred in critical cases.

2). **Post baking or hard baking.** In the hard baking, the developed wafer is kept in an oven at 120 °C for about 30 min. Post-baking or hard baking removes residual developing solvents and anneals the film to promote interfacial adhesion of the resist weakened by the developer penetration along the resist-substrate interface or by swelling of the resist (mainly for negative resists). Hard baking also improves the hardness of the film. Improved hardness increases the resistance of the resist to subsequent etching steps.

2.6.2. Photoresist

The resist is a photosensitive layer that is spun on to the wafer. The thickness of the resist is determined by its viscosity and the fining speed. Two types of resist – positive resist and negative resist. The resist becomes soft when exposed to UV light (+ve resist) or hard (-ve resist). The Figure 2.7 shows the pattern differences generated from the use of positive and negative resist.

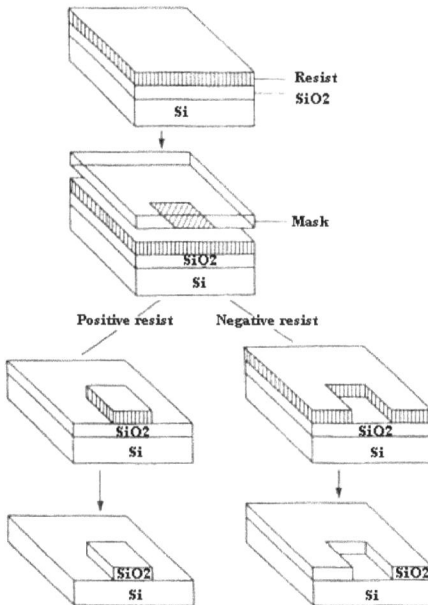

Figure 2.7. The pattern differences generated from the use of positive and negative resist.

2.6.3. Etching

Etching is classified into two broad areas: Dry Etching and Wet Etching.

2.6.3.1. Dry Etching

In order to pattern silicon into the extremely small electronic devices required for integrated circuits, plasma etching techniques were developed by the semiconductor electronics industry. Using plasma etching, highly directional etch profiles can be obtained, allowing many devices to be packed into a small surface area on a semiconductor substrate. Of the many dry etching techniques available, Reactive Ion Etching is the most commonly used one.

Reactive Ion Etching

In this technique, plasma is used to produce both chemically reactive species and ions. Etching takes place through the combined actions of energetic ions accelerated to the substrate, which damage the surface and remove material through sputtering, and through the chemical reactivity (perhaps increased by sputter damage) of the radicals produced in the plasma. RIE generally produces the fastest etch rates and allows for highly directional etches.

In many cases it is important to etch away one material while leaving other materials untouched. This is referred to as etch selectivity. Selectivity of etching processes is enhanced through the use of chemical reactants that only attack the layers to be etched. RIE can be highly selective by appropriate choice of etch gas.

2.6.3.2. Wet Etching

Wet etching of silicon is used for cleaning, shaping, polishing and characterizing structure and compositional features. In wet etching the material to be etched is dissolved when immersed in a chemical solution. Wet chemical etching provides higher degree of selectivity than dry etching techniques. Wet etching is often faster. It gives etch rate of a few microns to several tens of micron/minute for isotropic etchants whereas dry etching has about 1 micron/minute for an isotropic etchant. More recently though, with ECR dry etching, etch rates of up to 6 microns/minute were achieved. There are two types of wet etching: isotropic etching, anisotropic etching. A complete discussion of both etching is in Chapter 4.

2.7. Metallization

After the devices in the silicon substrate have been fabricated, they must be connected together to perform circuit functions. This process is called metallization, and is performed using one of several available vacuum deposition techniques.

To serve as an effective interconnect metallization on silicon, the metal chosen must meet all of the following requirements to at least a satisfactory level:

1) Low-resistance electrical contact to the silicon.

2) Limited reactivity with silicon for a stable contact.

3) High electrical conductivity so high current is easily carried without voltage drops.

4) Good adherence to the underlying silicon dioxide or other dielectric.

5) A pattern must be easily definable in the layer.

6) The deposition method must be compatible with already existing structures.

7) The metallization must uniformly cover steps in the surface topography.

8) The metallization must be able to withstand "electromigration." (Electromigration is the migration of the atoms in the metallization caused by the flow of current).

9) The metallization must not corrode under normal operating conditions.

10) It must be possible to bond easily to the metallization to allow external connection.

11) The metallization must be economically competitive.

No one metal perfectly meets all of these requirements. However, aluminium does meet all of these requirements quite well. Accordingly, aluminium is the metal most often chosen for device interconnection. Recent work has shown that the performance of aluminium can be improved upon by the introduction of small amounts of other elements. The tendency of aluminium to react with silicon can be halted by introducing a small percentage of silicon in the aluminium during deposition. In instances where aluminium does not meet the requirements of the metallization, multilayered structures are often utilized. Each layer will meet some of the requirements, and a combination of the layers will result in satisfying all of them.

There are several metallization techniques described as follows:

(1) Filament evaporation is the simplest and least expensive deposition method. However, contamination level of the deposited materials is often sufficiently high to interfere with the functioning of the device.

(2) Electron-beam evaporation (frequently called **E-beam**) uses a high-intensity focused beam of electrons to heat the material to be evaporated. Because only electrons come in contact with the material to be evaporated, it can be a low contamination process.

(3) Flash evaporation is similar to filament evaporation, in that the material is evaporated by thermal resistance heating, but the similarity ends here. Flash evaporation uses a continuously fed spool of wire (or in some cases, stream of pellets or powder) incident on a heated ceramic bar for the deposition. This deposition technique combines the speed and contamination-free features of E-beam deposition with the radiation-free feature of filament evaporation, and still offers the option of depositing composite layers.

(4) Induction evaporation is a recent innovation that offers some attractive features for certain coating problems. A radio-frequency source is used to couple powder into the metal to be evaporated in the crucible. The energy melts the metal, resulting in evaporation from certain regions. This method is not in common use in semiconductor operations.

(5) Sputtering is the last vacuum deposition method encountered in semiconductor processing. In sputtering, ions of inert gas are introduced into the chamber after a satisfactory vacuum level has been reached. An electric field ionizes these atoms, and causes them to move to one plate/electrode in the chamber called target. When the ions strike the target, they dislodge atoms from it, depositing them on the substrates facing the target. This process is shown in Figure 2.8. Sputtering can be accomplished using either DC voltage or RF voltage or DC magnetron or RF magnetron, and it can be used to deposit almost any material, although the deposition rate is often extremely low. Adhesion of layers deposited using this technique is generally very good. Argon is the inert gas used mostly in sputtering because of its relatively low cost and its good deposition rate.

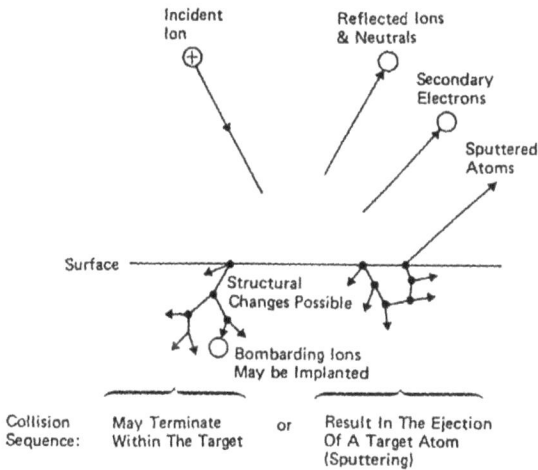

Figure 2.8. Sputtering of atoms from target.

2.8. Device Processing

The remaining processing steps that the devices undergo between metallization and final stage are much important as the initial steps. These steps are:

1) *Backside preparation;

2) Wafer short;

3) Device/die separation (Dicing).

After testing the devices on the wafer, the next processing step is to separate the devices into individual die for final packaging. There are many methods of separating the devices. One of them *dicing* is the most commonly used method.

References

[1]. S. A. Campbell, The Science and Engineering of Microelectronic Fabrication, *Oxford University Press,* New York, 1996.
[2]. S. Wolf and R. N. Tauber, Silicon Processing for the VLSI Era, Vol. I: Process Technology, *Lattice Press,* Second Ed., Sunset Beach, Ca, USA, 2000.
[3]. M. Madou, Fundamentals of Microfabrication, *CRC Press*, New York, 1997.
[4]. S. M. Sze, VLSI Technology, *Mc Graw-Hill*, New York, 1998.
[5]. K. Worhoff, Optimized LPCVD Si_xON_y waveguides covered with calixarene for non-critically phase-matched second harmonic generation, Ph. D thesis, *Univ. of Twente,* the Netherlands, May, 1996.
[6]. R. S. Rosler, Low pressure CVD production processes for poly, nitride and oxide, *Solid-State Technology*, 63, April, 1977, pp 63-70.
[7]. K. Watanabe, T. Tanigaki and S. Wakayama, The properties of LPCVD SiO_2 film deposited by SiH_2Cl_2 and N_2O mixtures, *Journal of Electrochemical Society,* 128, 1981, p. 2630.
[8]. W. Gleine and J. Müller, Low pressure chemical vapor deposition silicon-oxynitride films for integrated optics, *Appl. Opt.* 31, 1992, p. 2036.
[9]. E. Van de Ven, *Solid State Technology*, 161, April, 1981.
[10]. B. Reynes and J. C. Bruyere, PECVD silicon nitrides with low hydrogen content, *Sensors and Actuators A*, 33, 1992, pp. 25-28.
[11]. W. R. Knolle, J. W. Osenbach and A. Elia, Characterization of oxygen doped, plasma-deposited silicon nitride, *Journal of Electrochemical Society,* 135, 1988, p. 1211.
[12]. W. A. P. Claassen, W. G. J. N. Valkenburg, W. M. van de Wijgert and M. F. C. Willemsen, Influence of deposition temperature, gas pressure, gas phase composition, and RF frequency on composition and mechanical stress of plasma silicon nitride layers, *Journal of Electrochemical Society,* 132, 1985, p. 893.

Chapter 3

Thin Film Materials

Thin films of Silicon dioxide, Silicon nitride and Polysilicon have been utilized in the fabrication of absolute micro pressure sensor. These materials are studied and discussed in this chapter. Properties of polysilicon thin films are emphasized in detail.

3.1. Silicon Dioxide (SiO₂)

Silicon dioxide is commonly used as an insulator in integrated circuits. In MEMS it has been used as an electrically isolate component and as a structural material. Its basic properties are listed for reference in Table 3.1.

Table 3.1. Properties of silicon dioxide at room temperature.

Property	Value
Density	2.65 g/cm^3
Melting point	1728 °C
Young's modulus	66 GPa
Tensile strength	69 MPa
Thermal conductivity	$1.4 \times 10^{-2} \text{ w/ °C-cm}$
Dielectric constant	3.78
Resistivity	10^{12} Ω-cm
Energy gap	8 eV
Index of refraction	1.46
Thermal coefficient of expansion	$7 \times 10^{-6} \text{/°C}$

Silicon dioxide is a common component of glasses and is, as such, a very weak and brittle material. Thin films of oxide have a compressive internal stress of the order of 1 GPa. Despite this, due to the fact that silicon dioxide is less stiff than other thin film materials, it is used as a mechanical material in high sensitivity applications. Silicon dioxide, with its low thermal conductivity, is a natural thermal insulator, a property that has been exploited for the production of integrated thermal detectors. With a low tensile strength, silicon dioxide is susceptible to mechanical fracturing.

One major feature of silicon dioxide is its property as an insulator. With a band gap of 8 eV, silicon dioxide can effectively separate different layers of conductors with little electrical interference.

3.2. Silicon Nitride (Si₃N₄)

Silicon nitride is a material that is employed in a variety of applications. Since it does not react well with many etching solutions, silicon nitride is often used to prevent impurity diffusion and ionic contamination. Its basic properties are listed in Table 3.2.

Table 3.2. Properties of silicon nitride at room temperature.

Property	Value
Density	3.1 g/cm^3
Melting point	$1900 \,^{\circ}\text{C}$
Young's modulus	73 GPa
Fracture strength	460 MPa
Thermal conductivity	$0.28 \text{ W/cm-}^{\circ}\text{C}$
Dielectric constant	9.4
Resistivity	$10^{15} \, \Omega\text{-cm}$
Breakdown field	$1 \times 10^7 \text{ V/cm}$
Index of refraction	2.1
Coefficient of thermal expansion	$3 \times 10^{-5} \,/^{\circ}\text{C}$
Band gap	3.9- 4.1 eV

The silicon nitride films used in most MEMS devices are amorphous and are usually either sputtered or deposited by CVD techniques. These films are made with the following reaction, which occurs between 300-500 mT and 700-900 °C.

$$3\ SiH_2Cl_2\ (g) + 4\ NH_3\ (g) \longrightarrow Si_3N_4\ (s) + 6\ HCl\ (g) + 6\ H_2\ (g)$$

Simply adjusting the deposition temperature and the ratio of dichlorosilane (SiH_2Cl_2) to ammonia (NH_3) can control the stress of silicon nitride films. Silicon nitride has many mechanical properties that make it a desirable material to work with. It is a better thermal insulator than polysilicon, which can be important for isolating surface micro-machined structures. Also, its high mechanical strength makes it an ideal film for friction and dust barriers.

One of the unfortunate properties of silicon nitride is that it is not good insulator as silicon dioxide. With a bandgap 40 % smaller than SiO_2, the electrical isolation provided by silicon nitride is significantly less than that of silicon dioxide.

3.3. Polycrystalline Silicon (Polysilicon)

In applications involving surface micromachining, thin films of silicon are needed as a structural material. Since it is difficult to grow thin films of single crystal silicon, thin films of polycrystalline silicon are grown instead. These materials are now finding extensive use in the MEMS industry.

Polycrystalline silicon is also known as polysilicon and is composed of many small crystallites. In other words, it can be said that polysilicon is made up of many small single crystal domains called grains, whose orientations or alignments vary with respect to each other, i.e., they have random orientations. However, the most dominant orientation is <110>. The grain size in polycrystalline silicon is random and depends upon film thickness and the temperature at which it is deposited. Polysilicon is different from the single crystal silicon and amorphous silicon as shown in the Figure 3.1. In single crystal Si atoms are arranged in an orderly array that defines a periodic structure called lattice. In polycrystalline silicon there are many small grains, each having a well-organized structure, but differing from the neighbouring grains. In amorphous silicon atoms don't have a definite periodic arrangement of their atoms.

The fabrication of surface micro-machined MEMS devices utilizes polycrystalline Si as the structural material because it has mechanical properties that are comparable to single crystal Si (See Table 3.3).

Poly-Si is compatible with high temperature processing and interfaces very well with thermally deposited SiO_2.

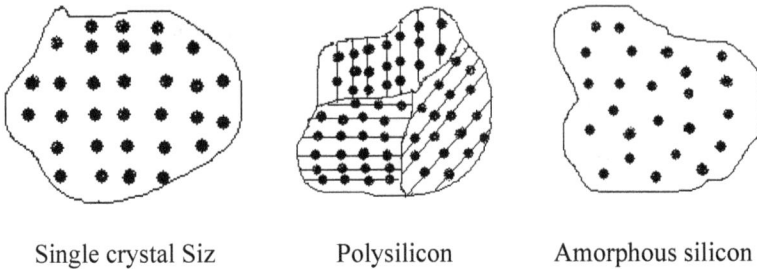

Single crystal Siz   Polysilicon   Amorphous silicon

Figure 3.1. Shows difference in structures of Single crystal Si, Polysilicon, and Amorphous silicon.

Table 3.3. Comparison of single crystal Si and Polysilicon material.

Material property	Single crystal Si	Polysilicon
Thermal conductivity (W/cm^0K)	1.57	0.34
Thermal expansion $(10^4/^0K)$	2.33	2-2.8
Specific heat (cal/g^0K)	0.169	0.169
Piezoresistance coefficients	Gauge factor of 90	Gauge factor of 30 (>50 with laser recrystallization)
Density (g/cm^3)	2.32	2.32
Fracture strength (GPa)	6	0.8 to 2.84 (undoped poly Si)
Dielectric constant	11.9	Sharp maxima of 4.2 and 3.4 eV at 295 and 365 nm respectively
Residual stress	None	Varies
Temperature resistivity coefficient (TCR), $(^0K^-)$	0.0017 (p-type)	0.0012 non linear, + or - through selective doping, increases with decreasing doping level, can be 0.
Poisson ratio	0.262 maximum for (111)	0.23
Young's modulas $(10^{11}N/m^2)$	1.90 (111)	1.61
Resistivity at room temperature (Ohm x cm)	Depends on doping $(2.3\times10^5$ Ohm x cm)	7.5×10^4 always higher than single crystal silicon

3.3.1. Properties of Polysilicon Film

3.3.1.1. Residual Stress

Polysilicon thin films are generally under a state of stress, commonly referred to as residual stress (Thin-film materials, which are attached to relatively thick substrates, can exhibit internal stress, which results from the formation or deposition of the thin film. Residual stress plays an extremely important role in MEMS and microelectronic devices. But it is not an intrinsic property; it depends on the specific substrate and the process used to form the film). In polysilicon micromechanical structures, the residual stress in the films can greatly affect the performance of the device. In general, deposited polysilicon films have compressive residual stresses. The highest compressive stresses are found in amorphous Si films and polysilicon films with a strong columnar (110) texture. For films with fine-grained microstructures, the stress tends to be tensile. For the same deposition conditions, thick polysilicon films tend to have lower residual stress values than thin films, especially true for films with a columnar microstructure. Annealing can be used to reduce the compressive stress in deposited polysilicon films.

Origins of Residual Stress in Polysilicon Thin Films:

1) Thermal expansion mismatch between a thin film and a substrate;

2) Chemical reactions far from equilibrium;

3) Thermal oxidation of silicon;

4) LPCVD of polysilicon and silicon nitride;

5) Non-ideal crystal structures;

6) Substitutional dopant incorporation;

7) Epitaxial mismatch;

8) Damage by ion-implantation;

9) Rapid deposition process (evaporation, sputtering).

These stress states in thin films are changed during the high temperature process steps (annealing).

For polysilicon films deposited at 650 °C, the compressive residual stress is typically on the order of 5×10^9 to 10×10^9 dyne/cm^2. However, these stresses can be reduced to less than 10^8 dyne/cm^2 by annealing the films at high temperature (1000 °C) in a N$_2$ ambient. Compressive stresses in fine grained polysilicon films deposited at 580 °C (100-Å grain size) can be reduced from 1.5×10^{10} to less than 10^8 dyne/cm^2 by annealing above 1000 °C. Rapid thermal annealing (RTA) is a fast and effective method of stress reduction in polysilicon films. For polysilicon films deposited at 620 °C with compressive stresses of about 340 MPa, a 10 seconds annealing at 1100 °C was sufficient to completely relieve the stress.

3.3.1.2. Young's Modulus

The effective Young's modulus of polysilicon, since it consists of dispersed crystallites, varies with film texture. The Young's modulus of polysilicon ranges from 140 to 210 GPa depending on crystal structure and orientation. Recent research has shown that the Young's modulus of polycrystalline films is highly dependent of deposition conditions. A polysilicon deposition and annealing process that yields a consistent Young's modulus is very desirable. The grain size in polysilicon films is typically a large fraction of the film thickness, and these films are technically considered as "multi-crystalline" films. The films exhibit preferential grain orientations that vary with temperature. Since an ideal film does not exhibit orientation dependence for its mechanical properties, a depositing film at 590 ^{0}C, which is the transition point between polycrystalline and amorphous silicon, is an effective method of producing an isotropic film of polysilicon. At this temperature the amorphous silicon will recrystalize during annealing, which produces films with a nearly uniform Young's modulus of 165 GPa [1].

3.3.1.3. Roughness

During fabrication of micro-machined devices, polysilicon films undergo one or more high temperature processing steps e.g., doping, annealing, and thermal oxidation. These high temperature steps can cause recrystallization of the polysilicon grains leading to a reorientation of the film and a significant increase in average grain size.

As result, the polysilicon surface roughness increases with the increase in grain size. The roughness often observed on polysilicon surfaces is due to the granular nature of polysilicon. The smooth surface is only obtained by depositing the film in the amorphous phase followed by a subsequent crystallization or using the chemical mechanical polishing process that reduces surface roughness with minimal film removal.

3.3.1.4. Electrical Properties

The electrical properties of polysilicon depend strongly on the grain structure of the film. The grain boundaries provide a potential barrier to the moving charge carriers, thus affecting the conductivity of the films. The resistivity of the polysilicon film is influenced by its structure, which in turn depends on the deposition conditions of the film. The resistivity is lowest for low-pressure film deposited in an initially amorphous form and subsequently crystallized, consistent with larger grains in the films. The grain size increases as the film thickness increases, and the resistivity decreases.

The electrical characteristics of polysilicon thin film depend on the doping as in the case of single crystal silicon – heavier doping results in lower resistivity. But polysilicon is more resistive than single crystal silicon at any given level of doping. Common dopants for polysilicon include arsenic, phosphorus, and boron. Polysilicon is usually deposited undoped and introduced with the dopants later on after deposition.

There are three ways to dope polysilicon, namely diffusion, ion-implantation and in-situ doping. Diffusion doping consists of depositing a very heavily doped silicon glass over the undoped polysilicon. This glass serves as the source of dopant for the polysilicon. Dopant diffusion takes place at a high temperature, i.e. at 900 - 1000 °C. Ion-implantation is more precise in terms of dopant concentration control and consists of directly bombarding the polysilicon layer with high-energy ions. In-situ doping consists of adding dopant gases to the CVD reactant gases during the epi deposition process.

Control of the electrical conductivity of the polysilicon layer is carried out using doping of compatible specie such as boron for p-type. In the present work, the variation in sheet resistivity with increasing doping temperature is shown in Figure 3.2 for the doping of boron and phosphorous in polysilicon by diffusion technique taking into account the underneath layer over which the polysilicon film is deposited. Also, the variation in sheet resistivity with doping concentration of boron and phosphorous is shown in Figure 3.3. Thus, in order to design the doped polysilicon resistors of desired value; one can control sheet resistivity by varying the doping temperature and doping concentration. In order to measure the absolute pressure, a micro pressure sensor based on boron doped polysilicon piezoresistors has been fabricated [2].

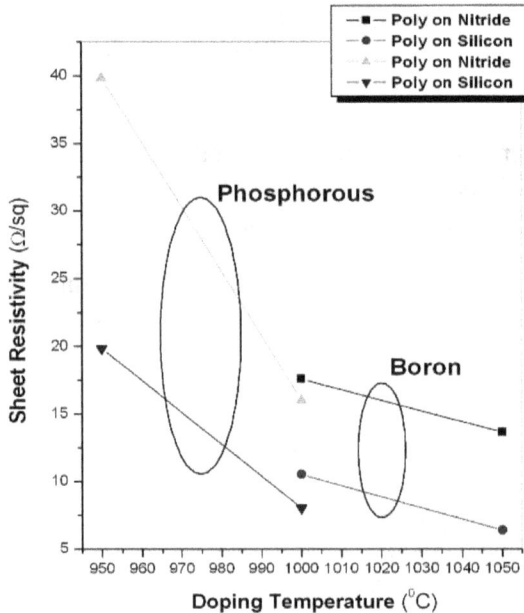

Figure 3.2. Variations of sheet resistivity with doping temperature.

3.3.2. Polysilicon as a Sacrificial Layer

Many methods have been developed to realize the membrane over a cavity structure of the pressure sensor. One approach includes the sacrificial layer deposition and etching by using surface micromachining technology [3-5]. In order to realize the square membrane over a conical cavity, polysilicon thin film can be used as a sacrificial layer.

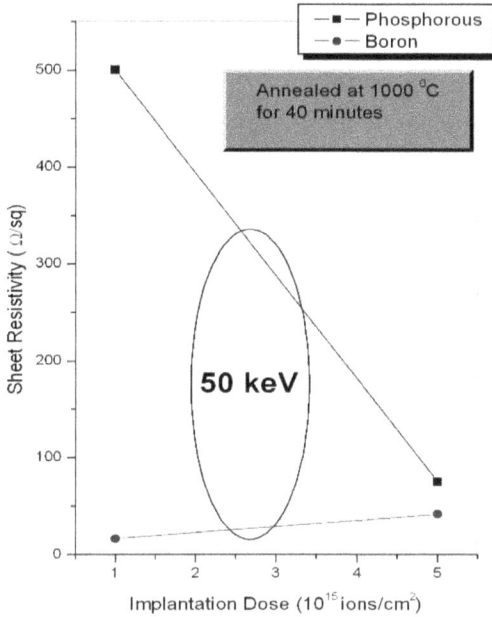

Figure 3.3. Variations of sheet resistivity with implanted dose.

In the present work, thin film of polysilicon has been used as a sacrificial layer deposited over the masking layer composed of silicon dioxide and silicon nitride above the silicon substrate for the purpose of membrane formation during fabrication of micro pressure sensor [6]. Because the etchant KOH has same etch rate for both polysilicon as well as silicon substrate in bulk.

3.3.3. Effect of Doping Temperature on Polysilicon Grains

Recent studies have analyzed the grain size of the polysilicon film and its impact on the device characteristics [7-9]. Thermal conductivity of polysilicon is another parameter, which carry importance due to its inherent association with the grain size of the polysilicon layer [10].

In the recent work, the thermally treated polysilicon film has been analyzed for its topological details using AFM in contact mode under ambient temperature and pressure. The grain size of the polysilicon film has been observed with varying temperatures during boron doping. The experimental detail is as follows.

Starting from a p-type (100) Si wafer, thermal oxide was grown to thickness of 0.5μm using a conventional dry-wet-dry procedure in a quartz furnace [11]. Silicon nitride was then deposited over the oxide using LPCVD technique, at 780 ^{0}C, to a thickness of 0.15 μm. The thickness of the composite layer was crucial to the strain balance at the interface of the silicon dioxide and silicon nitride [12]. Polysilicon was deposited on the silicon nitride bed using LPCVD at 620 ^{0}C in a furnace maintaining a silane (SiH$_4$) flow rate of 50 cm^3/min and a process pressure of 0.3 torr. The thickness of the polysilicon film was measured as 0.5 μm using a nanospec optical spectrophotometer. Boron doping of the polysilicon film was carried out in a furnace at four different temperatures of 920, 970, 1020, and 1070 ^{0}C, for 40 minutes in nitrogen ambient employing conventional flow rates [11]. The borosilicate glass (BSG), which is invariably deposited on the polysilicon, was etched out in buffered HF. The complete removal of the BSG is ensured prior to resistivity measurements and the AFM probing.

Atomic Force Microscopy (AFM) in contact mode was employed to analyze the grain size of the doped polysilicon layers. The AFM images of the polysilicon films doped typically at 920 and 1070 ^{0}C have been shown in the Figures 3.4 and 3.5.

The AFM micrographs confirm the polycrystalline nature of the doped film. The doping temperature does not affect the grain size of the polysilicon as is evident from Figures 3.4 and 3.5. Careful examining of the micrographs indicates ordering of the grains in circular form. The density of the grains in one arrangement grows further with increasing doping temperature. Figures 3.6 and 3.7 show the smaller area scan of the micrographs for polysilicon layers doped at 920 °C and 1070 °C respectively. In both the images, a size of 80 nm of the polysilicon grain confirms grain size invariance with increasing doping temperature. Variation of grain size of the polysilicon film with varying temperature has been considered a common phenomenon [9]. An undoped polysilicon film has shown monotonous increment of the grain size with increasing temperature. However in the present case, stability in the grain size with varying doping temperature during boron diffusion attracts attention. Ordering of the grains and the increasing grain density in each arrangement with increasing doping temperature can be seen in the Figures 3.6 and 3.7.

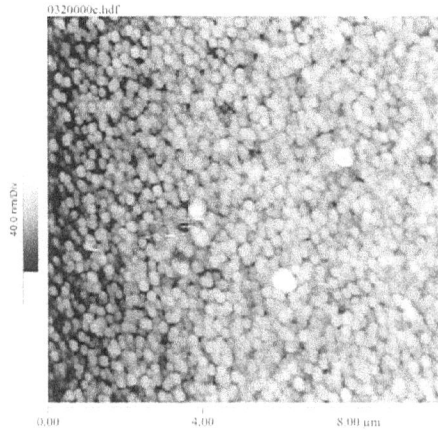

Figure 3.4. AFM image of polysilicon grains doped with boron at 920 °C (Scan area - 10 μm × 10 μm).

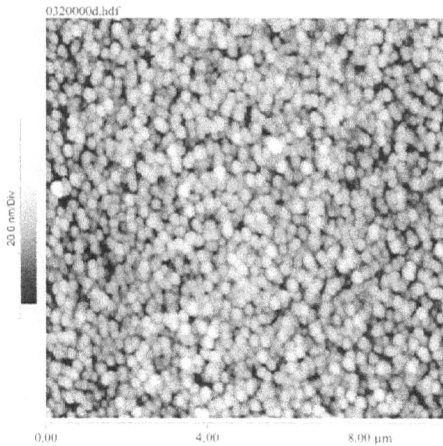

Figure 3.5. AFM image of polysilicon grains doped with boron at 1070 °C (Scan area – 10 μm × 10 μm).

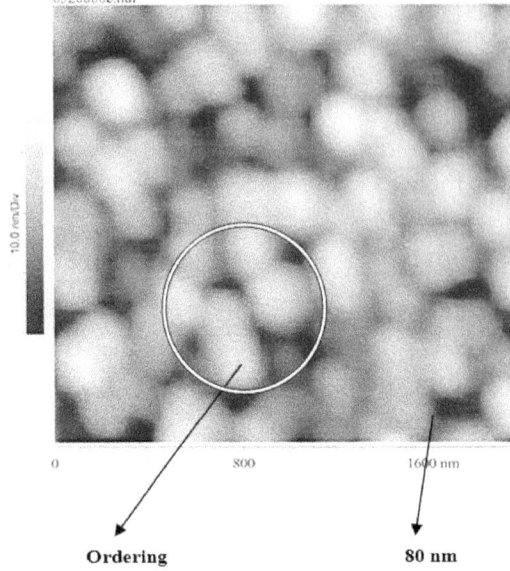

Figure 3.6. AFM image of polysilicon grains doped with boron at 920 °C
(Scan area – 2 μm × 2 μm).

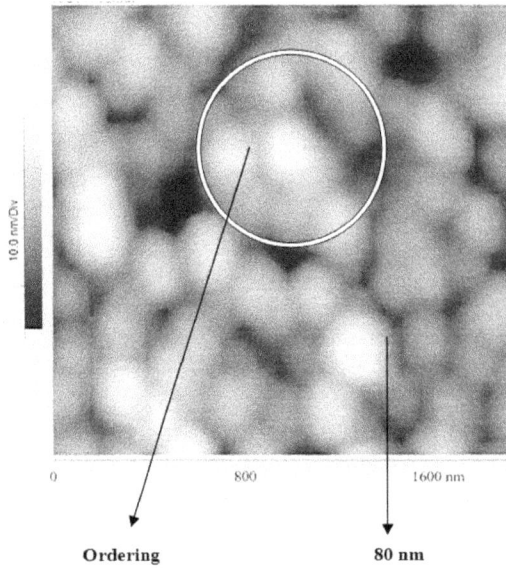

Figure 3.7. AFM image of polysilicon grains doped with boron at 1070 °C
(Scan area – 2 μm × 2 μm).

A less ordered arrangement of grains at 920 ^{0}C, as can be seen in Figure 3.6, has been changed into more ordered arrangement at 1070 ^{0}C, as shown in Figure 3.7.

The energy supplied by the increasing temperature seems to be utilized in the orderly arrangement of the grains. This might be due to the presence of boron doping during high temperature annealing of the polysilicon film. It has been found that the boron doping controls the electrical conductivity of the polysilicon film and the conductivity do not vary due to the size of the grains or clustering of the grains. Also, boron has a tendency to diffuse inside the grains [8]. The ordered arrangement of the polysilicon grains seems attractive as it leads to the nano scale entities, which could be realized by reducing the grain size during LPCVD. This is also associated with ordered arrangement of the grains, which grows with increasing doping temperature.

The present results show the experimental evidence of grain growth suppression, which is induced by boron thermal diffusion in polysilicon. The surface energy associated with individual grains and inward force due to boron doping seems responsible for the grain growth suppression.

Polycrystalline silicon grain size invariance with increasing temperature has been shown under the influence of boron thermal diffusion. The heat energy supplied due to increasing temperature results into dependent orderly arrangement of the grains with no change in the grain size that provides a new front in the area of nanotechnology.

3.3.4. Advantages and Disadvantages of Polysilicon Film

There are several advantages of polysilicon in general and of fine grained (FG) polysilicon in particular are:

1) High degree of thickness uniformity.

2) Polysilicon can be dielectrically isolated, eliminating junction leakage commonly present in ion-implanted, single crystal diaphragm pressure sensors.

3) Polysilicon can be deposited on a wide range of insulator-coated substrates.

4) The gauge factor of polysilicon is larger comparable to metal alloys.

5) Smooth surfaces and low defect densities.

6) Homogeneous, repeatable mechanical properties.

7) Better line width control than other, larger-grained films.

8) Nearly random grain orientations eliminate alignment concerns.

Disadvantages of polysilicon are:

1) Deposition on substrate with different coefficients of expansion can induce an in built strain or cause unwanted stress.

2) All physical properties and gauge factor depend on film morphology or structure and thus on the processing.

3) Technological problem can arises from polysilicon films being deposited under stress, which can result in rupture.

3.3.5. Applications of Polysilicon

Polysilicon is used in various applications in microelectronics and micromechanical devices such as sensors, accelerometers and actuators, because of its property to deposit on insulator substrate.

Applications of polysilicon films

1) As a gate electrode in MOSFET.

2) As base and emitter contact for bipolar transistor.

3) As first level local inter-connector.

4) High value resistors.

5) Diffusion sources for both active devices and contacts.

6) Used in thin film transistor, accelerometer, sensor and actuator.

7) Used as a sacrificial layer in MEMS technology.

Thin films of polycrystalline silicon are widely used as gate electrode in MOS transistor to enable further reduction in dimensions. These devices could operate at high temperature. It is also used for interconnection in MOS circuits. It is used as resistor, as well as in ensuring ohmic contacts for shallow junctions. When used as gate electrode, a metal (such as tungsten) may be deposited over it to enhance its conductivity.

As a gate electrode, it has also been proven to be more reliable than Al. It can also be deposited conformally over steep topography. Heavily doped poly thin films can also be used in emitter structures in bipolar circuits. Lightly doped poly films can also be used as resistors.

Polysilicon resistor on oxidized silicon exhibits excellent mechanical properties of silicon with the efficient insulation, which improves stability and high temperature operation. Polysilicon based piezoresistive pressure sensors present low nonlinearity, a very linear thermal drift and high stability.

References

[1]. B. Stark, Material properties, (also available online at - http://parts.jpl.nasa.gov/docs/JPL%20PUB%2099-1D.pdf

[2]. P. A. Alvi, K. M. Lal, V. P. Deshwal and J. Akhtar, Vacuum sealed cavity absolute micro pressure sensor employing polysilicon piezoresistors, in *Proceedings of the National Conference on Sensors,* Thapar Institute of Engineering and Technology, Patiala, (Punjab) India, 25-26 November 2005, pp. 21-25.

[3]. J. T. Kung and H. Lee, An integrated air-gap-capacitor pressure sensor and digital readout with sub-100 attofarad resolution, *Journal of Microelectromechanical System,* Vol. 1, 1992, pp. 121-129.

[4]. S. Guo, J. Guo, and W. H. Ko, A monolithically integrated surface micromachined touch mode capacitive pressure sensor, *Sensors and Actuators A, Phys.,* Vol. A80, 2000, pp.224-232.

[5]. H. K. Trieu, N. Kordas, and W. Mokwa, Fully CMOS compatible capacitive differential pressure sensor with on chip programmabilities and temperature compensation, in *Proceedings of the IEEE International Conference on Sensors, 2002,* pp. 1451-1455.

[6]. P. A. Alvi, B. D. Lourembam, V. P. Deshwal, B. C. Joshi and J. Akhtar, A process to fabricate micro-membrane of Si_3N_4 and SiO_2 using front side lateral etching technology, *Sensor Review*, Vol. 26, No. 3, 2006, pp. 179-185.

[7]. S. Kalainathan, R. Dhanasekaran, and P. Ramasamy, Grain size and size distribution in heavily phosphorous doped polycrystalline silicon, *Journal of Growth,* 104, 1990, p. 250.

[8]. G. Franco, M. Priulla, G. Renna, and G. Scerra, Influence of the polysilicon doping on the electrical quality of thin oxides: a confrontation between vertical and horizontal furnaces, *Material Science in Semiconductor Processing,* 4, 2001, p. 153.

[9]. V. K. Sooraj and K. H. Miltiadis, Lateral polysilicon p^+-p-n^+ and p^+-n-n^+ diodes, *Solid-State Electronics*, Vol. 47, 2003, p. 653-659.

[10]. D. McConnel Angela, Uma Srinivasana, E. Goodson Kenneth, Thermal conductivity of doped polysilicon layers, IEEE/ASME, *Journal of Microelectromechanical Systems,* 10, 2001, p. 360-369.

[11]. C. Y. Chang and S. M. Sze, ULSI Technology, *McGraw-Hill,* New York, 1996.

[12]. Fariborz Maseeh and Stephen D. Senturia, Plastic deformation of highly doped silicon, *Sensors and Actuators,* A21-A23, 1990, p. 861-865.

Chapter 4

Anisotrpic Etching

4.1. Introduction

Anisotropic etching of silicon refers to the direction-dependent etching of silicon, usually by alkaline etchants like aqueous KOH, TMAH and other alkaline hydroxides like NaOH and LiOH. Due to the strong dependence of the etch rate on crystal direction and on etchant concentration, a large variety of silicon structures can be fabricated in a highly controllable and reproducible manner. Hence, anisotropic etching of silicon using aqueous KOH solution has been used widely and for long to easily fabricate a variety of devices and 3-D MEMS structures at a low cost. These include V-grooves for VMOS transistors, small holes for ink jets and diaphragms for MEMS pressure sensors. The actual reaction mechanism has not been fully understood for long and a comprehensive physical model for the process has not yet been developed. With increasing numbers of MEMS applications, interest has grown for process modeling, simulation and software tools useful for prediction of etched surface profile.

Chemical etching of silicon depends on crystal orientation, temperature and concentration of the etchant. Geometry of the area to be etched also influences the etch rate owing to the different crystal planes encountered during the etching process. In order to prepare an accurate physical model, experimental data under varying conditions are required. Therefore, anisotropic etching of (100) silicon has been carried out at varying temperatures and concentrations at CEERI, Pilani (CSIR, INDIA). In order to minimize the influence of other chemicals on the etching mechanism and therefore obtain more accurate results, pure KOH solution has been preferred over a number of mixtures with moderators like Ethylene Diamine Pyrocatechol (EDP) and Isopropyl Alcohol (IPA). Also, since the boiling point of a moderator like IPA is just 82.5 °C, the use of pure KOH solution enables the temperature of the etch solution to be raised up to its boiling point. The results obtained by experimentation have been used in order to understand and postulate a physical model for anisotropic etching.

During etching, bubbles of hydrogen gas are generated as a by-product of the reaction between Si and KOH. The quality of the surface generated after etching is largely determined by the size and properties of these bubbles. This is because when the hydrogen bubbles adhere to the silicon surface, they act as temporary localized etch masks. This masking effect tends to produce hillocks bounded by four (111) facets when etching of (100) is carried out. The bubbles tend to adhere more strongly to a silicon surface that is hydrophobic. It has been observed that the surface roughness depends on the hydrophilic or hydrophobic nature of the silicon surface. This in turn depends upon the KOH solution concentration. Hence, the data gathered at different KOH concentrations and temperatures for etch rate is also useful to determine surface quality. Surface quality was determined using Atomic Force Microscopy (AFM) on the etched surface samples.

In this chapter, the experimental data, results, relevant details of the experiment and theoretical background have been discussed.

4.2. Wet Etching Fundamentals

4.2.1. Isotropic Etching

Wet etching of silicon is used for cleaning, shaping, polishing and characterizing structure and compositional features. Wet chemical etching provides higher degree of selectivity than dry etching techniques. Wet etching is often faster. More recently though, with ECR dry etching, etch rates of up to 6 microns /minute were achieved. Modification of wet etchant and /or temperature can alter the selectivity and specially when using alkaline etchants to crystallographic orientations. Etching proceeds by reactant transport to the surface.

Isotropic etchants, also polishing etchants, etch in all crystallographic directions at the same rate; they usually are acidic, such as $HF/HNO_3/CH_3COOH$ (HNA), and lead to rounded isotropic features in Si. They are used at room temperature slightly above (<500 °C). Some alkaline chemicals etch anisotropically, i.e., they etch away crystalline silicon at different rates depending on the orientation of the exposed crystal plane.

Uses of Isotropic Etchants: When etching silicon with aggressive acidic etchants, rounded isotropic patterns form. The method is widely used for:

- Removal of work damaged surfaces;

- Rounding of sharp anisotropically etched corners (to avoid stress concentration);

- Removing of roughness after dry or anisotropic etching;

- Creating structures or planner surfaces in single crystal slices (thinning);

- Patterning single crystal, polycrystalline, or amorphous films;

- Delineation of electrical junctions and defect evaluation (with preferential isotropic etchants).

For isotropic etching of silicon, the most commonly used etchants are mixtures of nitric acid and hydrofluoric acid.

4.2.2. Anisotropic Etching

Anisotropic etchants etch desired structures in crystalline materials when carried out properly. Anisotropic etching results in geometric shapes bounded by perfectly define crystallographic planes since the rate of etching is direction dependent. A wide variety of etchants have been used for anisotropic etching of silicon, including aqueous solutions of KOH, NaOH, LiOH, CsOH, NH_4OH and quaternary NH_4OH with the possible addition of alcohol. Alkaline organics such as ethylenediamine, choline or hydrazine with additives such as pyrocatechol and pyrazine are employed as well.

4.2.3. Wet Anisotropic Etching Using Aqueous KOH Solution

The existence of selective and anisotropic etching for Si and SiO_2 has formed the basis of many of the initial micromachining investigations. There are two classes of etchants: the KOH-based etchants and the ethylene diamine based etchants. Both types show highly anisotropic behavior; etching <111> Si planes at a much slower rate than other planes. Both etchants show a doping dependence, with etch rates slowing considerably as the Si becomes more than 3×10^{19} cm^{-3} p-type with boron doping.

In order to calculate the angle between two crystallographic planes with known Miller Indices in case of a cubic lattice as of silicon, following well-known relation is used:

$$Cos(h_1k_1l_1):(h_2k_2l_2)=(h_1h_2+k_1k_2+l_1l_2).\{h_1^2+k_1^2+l_1^2).(h_2^2+k_2^2+l_2^2\}^{-1/2}$$

In case of <100> and <111> planes, the above equation modified as follows:

$$Cos<100> : <111> = 1/\sqrt{3} = 0.5774$$

$$<100> : <111> = Cos^{-1} (0.5774)= 54.74°$$

Thus a V-groove will be formed as a result of wet etching of Si<100> making an angle of 54.74° between the <100> and the<111> plane. (The angle between the planes is an important parameter for the necessary calculations for designing the mask layout). The making of angle between the planes <100> and <111> is shown in the Figure 4.1.

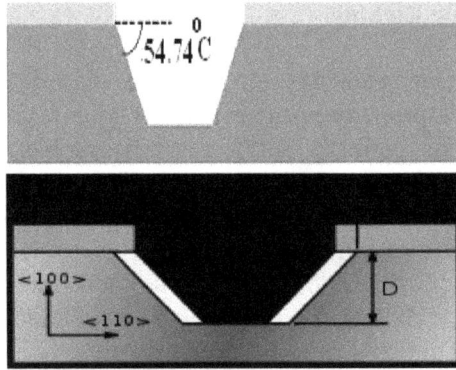

Figure 4.1. Illustration of the angle formed between the (100) and (111) planes after anisotropic etching.

According to [1], the overall etching reaction in KOH solution may be given by the equation:

$$Si + 4H_2O \longrightarrow Si(OH)_4 + 2H_2$$

It is the H_2 produced in this reaction that is seen as bubbles escaping from the reaction vessel. The Orthosilicic acid {$Si(OH)_4$} produced is soluble in the solution.

KOH is one of the most commonly used silicon etch chemistry for micromachining of silicon wafers. Ethylene diamine – based etchants (e.g. ethylene diamine-pyrocatechol, EDP) are a potential carcinogenic material so not commonly used. KOH etching depends not only on the crystallographic orientations of the wafers but also on the operating temperature and its concentration [2]. KOH etching is usually performed at 75 °C to 80 °C, though the etch rate is slower. This because at higher temperature (above 120 °C) it is found that Si_3N_4 layer which is used as a mask also starts etching. Also at higher temperature near boiling point due to the aggressive etching process delicate features in the device may get damaged.

4.3. Experimental Methodology

In order to gather the data necessary to be able to postulate a model for etching of silicon by KOH, several samples of (100) silicon wafer were prepared. Silicon (100) samples have been prepared starting with device grade chemical cleaning of the wafer. Thermal oxidation of 0.5-micrometer thickness was grown using wet-dry-wet sequence at 1100 °C, in a horizontal quartz furnace. LPCVD silicon nitride was deposited over that layer, at 780 °C for a thickness of 0.15 micrometer in a horizontal quartz furnace reactor. The combination of silicon dioxide and silicon nitride provides stable masking action in KOH solution at elevated temperatures. Square windows of 2.0 mm size were delineated in an array using photolithography. Dry etching was used to selectively remove silicon dioxide and silicon nitride. The samples were used for KOH etching in a reflux condenser equipped with a magnetic stirrer to maintain KOH concentration, as shown in Figure 4.2.

The wafers were etched at various temperatures and concentrations of KOH for a period of one hour each. No additives other than DI water for dilution were used in all experimentation in order to be able to develop a model based on data uninfluenced by the reactivity of other chemicals. Also, data could be gathered at higher temperatures since the addition of low BP impurities like isopropanol (B.P. 82.5 °C) has been avoided. The etch depth was measured by an indirect method using an optical microscope. This method makes use of the fact that the angle between the (100) and (111) planes in a silicon crystal is always 54.74°.

Figure 4.2. Reaction container for KOH etching at CEERI, Pilani (CSIR, India).

With the help of the scale provided with the optical microscope lens, the length x was measured under 50X magnification. The actual etch depth y was then estimated using the relationship: $y = x \tan54.74°$ (See Figure 4.3). Etch rate was then evaluated as the etched depth per unit time (μm/min).

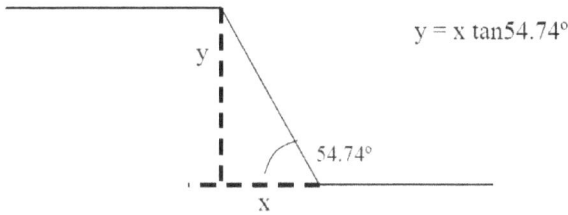

$$y = x \tan 54.74°$$

Figure 4.3. Etched geometrical view.

4.4. Surface Roughness due to KOH Etching

The origin of pyramid formation during etching may be attributed to the adhesion of H_2 bubbles to the silicon surface, as shown in Figure 4.4. The purity of the etchant and of the water used is also crucial for achieving smooth surfaces. Surface roughening has also been attributed to the etching conditions and surface inhomogeneities. Poor bubble detachment from the surface results in the formation of micro-pyramids and pits, as shown in Figure 4.4.

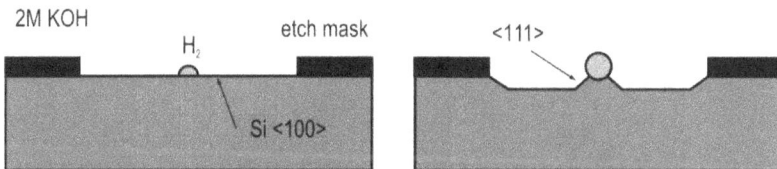

Figure 4.4. Schematic representation of hydrogen bubble formation on a masked Si (100) surface (left), and the resulting hillock formed by H_2 attachment, leading to a truncated pyramid (right).

Etching under ultrasound conditions in the presence of additives usually employed in micromachining baths leads to smooth and defect-free surfaces. The significant improvement in surface finish under these conditions indicates that hydrogen bubbles, which are temporarily attached during the etch process, are one of the main origins of surface roughness [3]. The AFM image of the sample surface shown in Figure 4.5 is further shown in Figure 4.6, but under ultrasound condition.

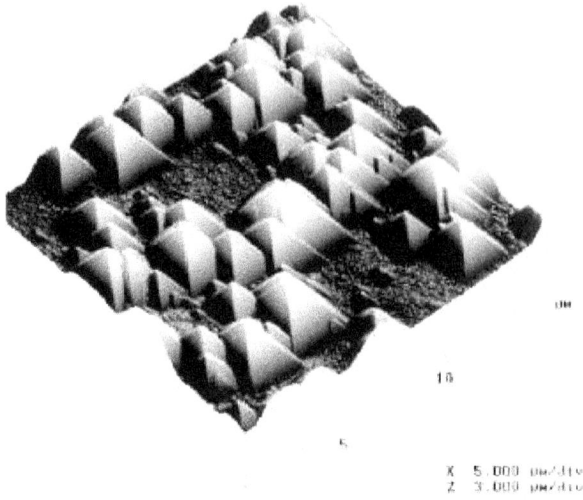

Figure 4.5. AFM image of a Si (100) surface etched in KOH solution
at 600 °C for 60 min.

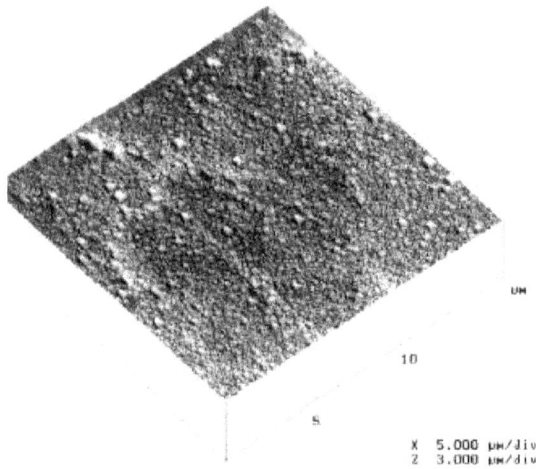

Figure 4.6. AFM image of a Si (100) surface etched in KOH solution
at 60 °C for 60 minutes but under ultrasound condition.

The formation of large hydrogen gas bubbles depends on the adhesion
between the KOH solution and the silicon wafer. If the surface is

hydrophilic, only small bubbles are formed and roughness is reduced and if the surface is hydrophobic, large bubbles and hence large hillocks are formed as shown in Figure 4.7. Concentrated KOH solutions tend to adhere more to Si surfaces, resulting in smaller bubbles and less roughness. These bubbles cause temporary localized etch stops equivalent to etch masks. Surface roughening in the form of hillocks can be observed and the resulting morphology is governed by the anisotropic etch rates producing pyramids made of four {111} facets.

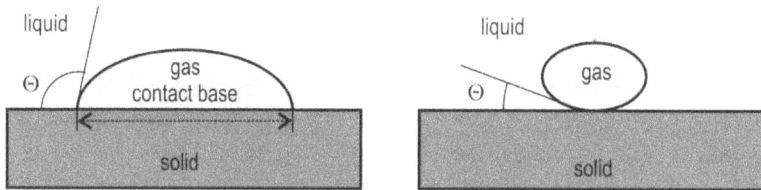

Figure 4.7. Schematic representation of hydrogen bubble formation for hydrophobic (left), and hydrophilic (right) silicon surfaces.

After studying the AFM images of the etched surfaces, shown in Figure 4.8, which are etched at 10 wt % and 20 wt % concentration of KOH at the same temperature (80 °C), we conclude that with increasing KOH concentration the surface quality improves (roughness decreases). This would suggest that at higher KOH concentrations, the silicon surface is more hydrophilic. This may be understood from the fact that due to the higher KOH concentration, the Si atoms at the crystal surface are OH⁻ terminated instead of being H terminated. Silicon surface, that is OH⁻ terminated, is more hydrophilic due to the presence of charge on the OH⁻ radical and the fact that water is a polar solvent.

The significant improvement in surface finish under ultrasound conditions indicates that hydrogen bubbles, which are temporarily attached during the etch process, are one of the main origins of surface roughness. Improvement in surface finish on application of a positive potential also points to OH⁻ termination of the surface. The positive potential attracts negatively charged OH⁻ ions and leads to greater Si-OH termination. This in turn increases the hydrophilic nature of the surface, favoring faster bubble detachment and improved surface finish.

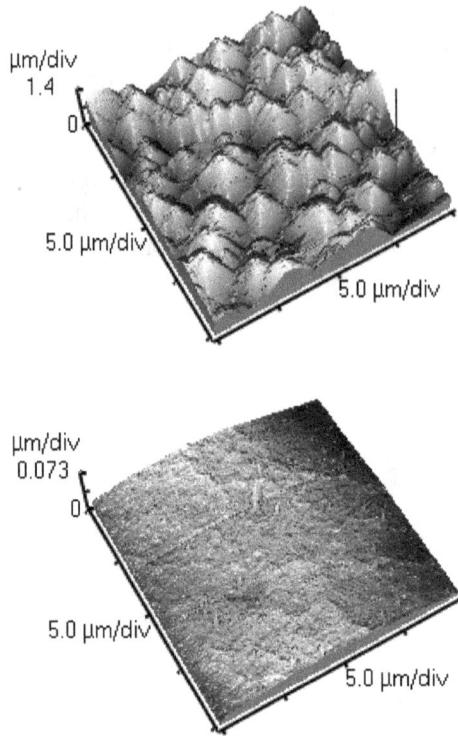

Figure 4.8. AFM image of etched silicon surface at 10 wt % concentration (left) and at 20 wt % concentration (right) of KOH at 80 °C.

4.5. Physical Models for KOH Etching

According to [4], it is believed that the etch rate is related not only to the macroscopic process, but also to the microscopic ones, which are denoted here by the temporary states. According to the above chemical process, for an atom on a {1 0 0} plane, the possible microscopic states during the etching process are shown in the Figure 4.9. To distinguish them clearly, digits have been marked on the corresponding microscopic state in the Figure 4.9.

With the macroscopic reaction, microscopic states will be transferred into other states. The transfer relation of the microscopic states is shown in Figure 4.10. It is known that the chemical reaction is a dynamic process and the chemical balance is a dynamic equilibrium.

From the macroscopic viewpoint, for the atoms participating in the reaction, when the number of the reacted atoms is equal to that of the produced ones, the reaction will reach the dynamic balance. It is just the same for the microscopic states. For one microscopic state, when the produced number is equal to the reacted number, the dynamic equilibrium will appear.

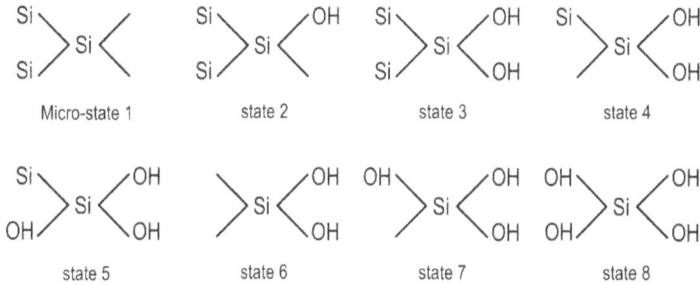

Figure 4.9. Diagram of possible existing microscopic states for {1 0 0} crystal planes during anisotropic etching in KOH.

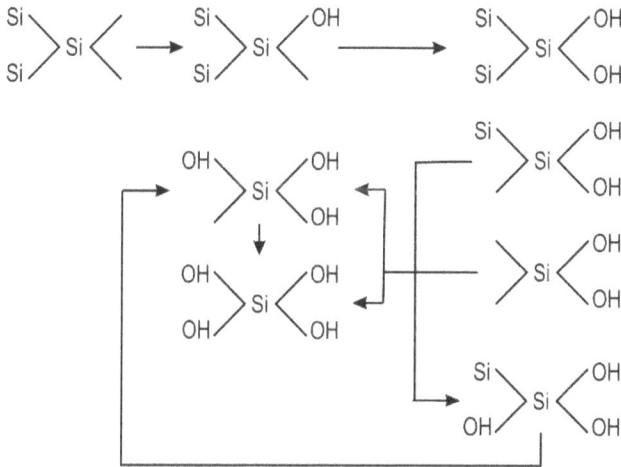

Figure 4.10. Illustration of the possible transfer relation of the states listed in Figure 4.9.

Because the etch rate is related to time, a main equation which is widely used in chemico-physics is adopted here. The main equation related to the transfer probability between different states is a kind of differential equation. It is distinctive and convenient to describe the chemical reaction using this equation. The important parameter in the main equation is the transfer probability. Once it is fixed, the main equation will be listed.

The microscopic states listed in the Figure 4.9 may be transferred into other states. The transform is not arbitrary, certainly. It follows the macroscopic reaction. For example, for state 6, it should transfer into orthosilicic acid $Si(OH)_4$. A hydroxyl ion can be combined into state 6, so it will transfer into state 7. At the same time, two hydroxyl ions can be combined with state 6, too. So, state 6 can be changed into state 8. The two transforms can both occur and their possibilities will determine the possible result. Corresponding to the microscopic states listed in the Figure 4.9, the possible transfer relation of the states is illustrated in the Figure 4.10, in which the transform possibilities are shown using arrows. The number of the microscopic state during the reaction is assumed to be n_i, where i stands for the marked number shown in the Figure 4.9. The transfer probability from state i to state j is P_{ij}. The main equations that describe the reactions shown in the Figure 4.10 are listed below:

$$\frac{dn_1}{dt} = -P_{12}n_1 + P_{21}n_2$$

$$\frac{dn_2}{dt} = P_{12}n_1 + P_{32}n_3 - P_{21}n_2 - P_{23}n_2$$

$$\frac{dn_3}{dt} = P_{23}n_2 + P_{43}n_4 + P_{53}n_5 + P_{63}n_6 - P_{32}n_3$$
$$- P_{34}n_3 - P_{35}n_3 - P_{36}n_3$$

$$\frac{dn_4}{dt} = P_{34}n_3 + P_{54}n_5 + P_{64}n_6 - P_{43}n_4 - P_{45}n_4 - P_{46}n_4$$

$$\frac{dn_5}{dt} = P_{45}n_4 + P_{35}n_3 + P_{75}n_7 - P_{54}n_5 - P_{53}n_5 - P_{57}n_5$$

$$\frac{dn_6}{dt} = P_{36}n_3 + P_{46}n_4 + P_{76}n_7 + P_{86}n_8 - P_{63}n_6$$
$$- P_{64}n_6 - P_{67}n_6 - P_{68}n_6$$

$$\frac{dn_7}{dt} = P_{87}n_8 + P_{57}n_5 + P_{67}n_6 - P_{78}n_7 - P_{75}n_7 - P_{76}n_7$$

$$\frac{dn_8}{dt} = P_{78}n_7 + P_{68}n_6 - P_{87}n_8 - P_{86}n_8.$$

The transfer probability P_{ij} obeys the Boltzmann distribution:

$$P_{ij} = \lambda \exp\left(\frac{E_i - E_j}{KT}\right),$$

where E_i and E_j stand for the microscopic activation energies of states i and j, respectively, λ is the vibrational frequency of the crystal lattice, K is the Boltzmann constant and T is the temperature [5]. Once the activation energy is fixed, the above equation will be dissolved.

4.6. Results and Inferences

The experimental results for the etch rate of KOH in silicon substrate varying with etching temperatures and etchant concentration have been shown in the Figures 4.11 and 4.12.

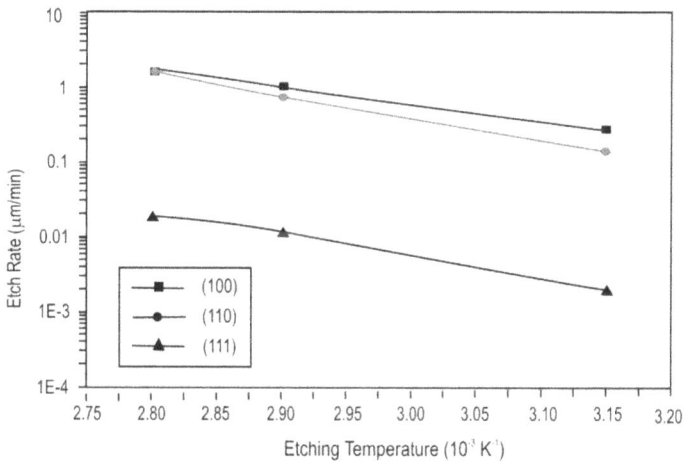

Figure 4.11. Shows variation in etch rate (experimentally) with increasing temperature.

The calculated results for the etch rate of KOH in silicon substrate varying with etching temperatures and etchant concentration have been shown in the Figures 4.13 and 4.14.

The high degree of correlation between the expected values obtained by calculations (based on the model proposed) and the experimentally observed results validates the model.

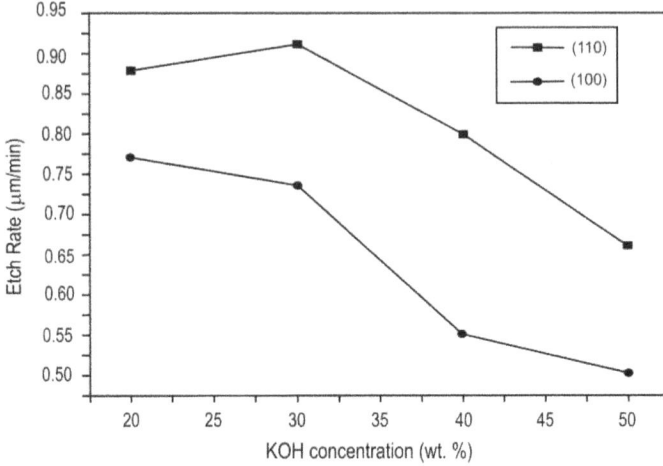

Figure 4.12. Shows variation in etch rate (experimentally) with KOH concentration.

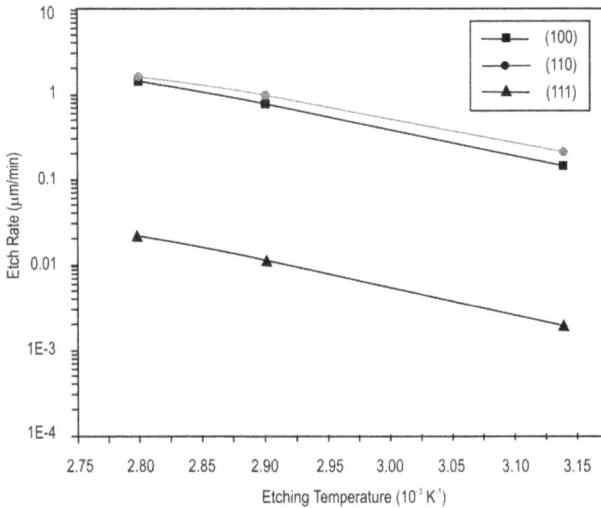

Figure 4.13. Shows variation in etch rate (by calculation) with increasing temperature.

In the experiments conducted at CEERI (CSIR, INDIA), the Arrhenius relationship postulated in the model proposed in [2, 4, 5] has been verified. This is seen clearly in the graph shown in Figure 4.15, for etching using 30- wt % KOH solution.

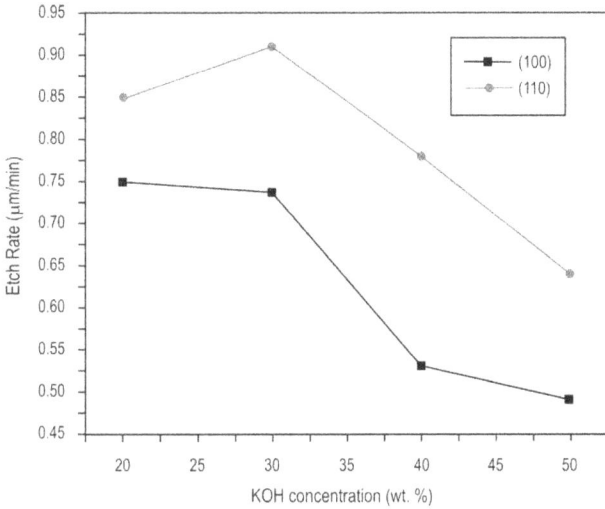

Figure 4.14. Shows variation in etch rate (by calculation) with etchant concentration.

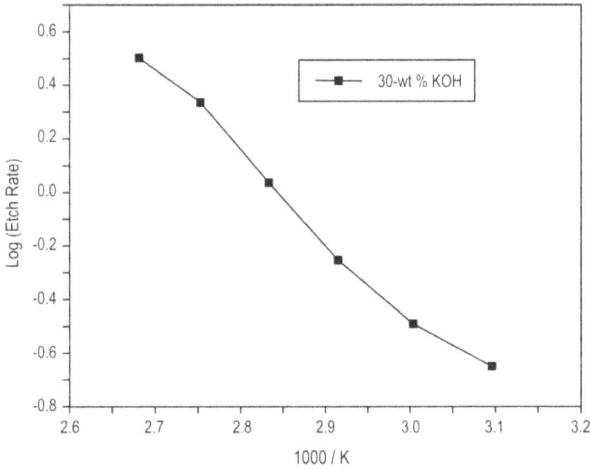

Figure 4.15. Shows variation of etch rate with temperature showing Arrhenius relationship.

95

Figure 4.16 shows the result obtained by experimentation at CEERI for the variation of etch rate of (100) Si with concentration at the temperatures of 50 °C, 60 °C and 70 °C. Concentration of the KOH solution was varied by adding the required amount of DI water to a standard 45 wt % solution.

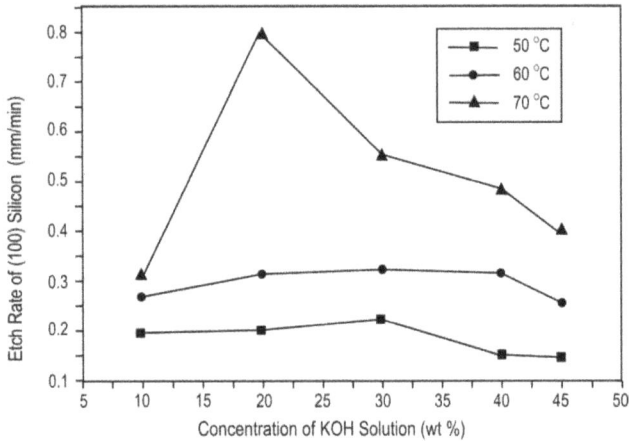

Figure 4.16. Shows variation of etch rate with concentration at temperatures of 50 °C, 60 °C and 70 °C.

The graph shown in Figure 4.16 indicates that variation of etch rate with concentration is less pronounced at lower temperatures like 50 °C and 60 °C than at 70 °C, since the etch rate itself at these temperatures is less. The variation in etch rate at 70 °C shows first an increase with concentration and then a decrease. An explanation provided by research conducted much earlier suggested a model where both OH⁻ and H_2O participate in the reaction with Si as follows:

$$Si + 2H_2O + 2OH^- \longrightarrow Si(OH)_4 + H_2$$

Such a reaction would proceed at a lower rate in the absence of any one of the reactants H_2O and OH⁻. Since at low KOH concentration, OH⁻ is in low concentration, and at high KOH concentration, H_2O is available in low concentration; the reaction would slow down at both high and low KOH concentrations. However, this is not a comprehensive explanation capable of explaining all the experimental results obtained.

Figure 4.17 shows the result obtained by experimentation at CEERI for the variations in etch rate with temperature for 45-wt % KOH solution.

A steep increase in etch rate is seen with temperature. Plots of etch rate versus 1/T (where T is absolute temperature) have shown a straight-line Arrhenius relationship between etch rate and temperature in Figure 4.15.

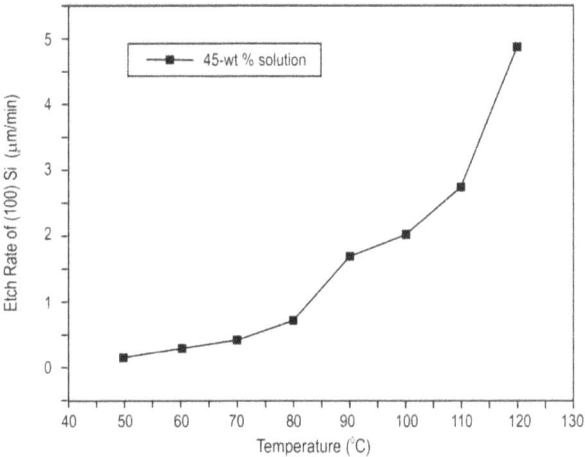

Figure 4.17. Shows variation of etch rate with temperature for 45-wt % KOH solution.

Figure 4.18 shows the result obtained by experimentation at CEERI for the variation of etch rate of (100) Si with concentration at the temperatures of 80 °C, 90 °C and 100 °C. The boiling point of the KOH solution increases almost linearly with concentration. The boiling point sets the upper limit of the maximum temperature allowed for a given concentration.

The etching rate shows a peak for each temperature curve that is shifted towards lower concentration with increasing temperature. At 80 °C, a maximum etching rate of 1.5 µm/min is obtained for 25 wt % KOH solution. For temperatures above 80 °C, the peak of the maximum etch rate is shifted to 15 wt % KOH solution. At 100 ^{0}C the maximum etch rate is about 3.0 times of that at 80 °C.

From the observations taken at CEERI (CSIR, INDIA), faster etching rate with smooth silicon surface has been observed at higher temperatures in low KOH concentration solutions. Higher KOH concentrations improve etched surface finish. Etch rate for a particular concentration increases with temperature.

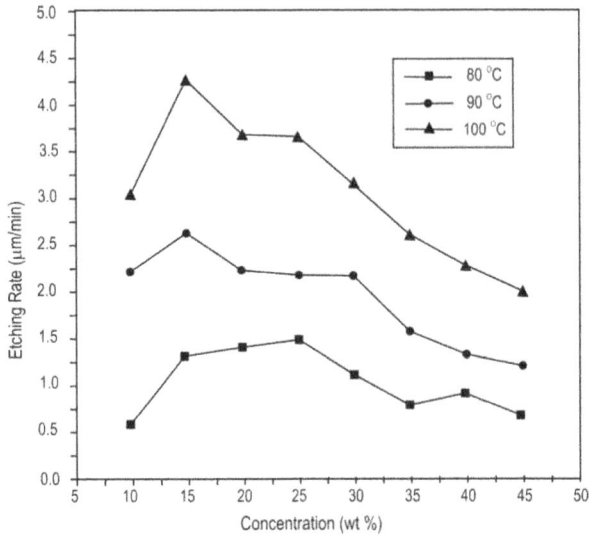

Figure 4.18. Variation in etch rate with concentration at temperatures of 80 °C, 90 °C and 100 °C.

References

[1]. Gad El-Hak, Mohammed, MEMS Handbook, *CRC Press*, 2000.

[2]. Seidel, H. et al, Anisotropic etching of crystalline silicon in alkaline solution, *Journal of the Electrochemical Society*, Vol. 137, No. 11, November, 1990.

[3]. Baum, Theo and David J Schiffrin, AFM study of surface finish improvement by ultrasound in the anisotropic etching of Si <100> in KOH for micromachining applications, *Journal of Micromechanics and Microengineering*, 7, 1997, pp. 338-342.

[4]. Jiang, Yanfeng and Qing-an Huang, A physical model for silicon anisotropic chemical etching, Institute of Physics Publishing, *Semiconductor Science & Technology*, 20, 2005, pp. 524-531.

[5]. Lysko, M. Jan, Anisotropic etching of the silicon crystal – surface free energy model, *Material Science in Semiconductor Processing*, 6, 2003, pp. 235-241.

Chapter 5

Fundamental Theory and Design of Pressure Sensor

Pressure sensor fabricated in this work is based on the piezoresistors. These piezoresistors undergo a change in resistance due to the applied pressure. The resistance change is converted into voltage signal by means of a Wheatstone bridge and by applying known pressures on the diaphragm and recording the output voltages the sensor can be calibrated.

A piezoresistive pressure sensor consists of three major active components namely, a diaphragm, Wheatstone bridge and the piezoresistors. These three components, responsible for the sensitivity of the device, are described in the following.

5.1. Stress Analysis for Thin Diaphragm

Understanding the deflection behavior of micro-machined diaphragms is necessary for the designing mechanical sensors such as pressure sensors and accelerometers. Several papers have recently been published regarding the deflection characteristics of the micro-machined diaphragms [1-7]. For these sensors, the applied load is assumed to be constant over the diaphragm surface. Small deflection theory is inadequate for describing diaphragm behavior. Large deflection theory is better, but generally does not consider built in stress effects.

The large deflection problems for square diaphragm with clamped edges have been solved by different methods [8-11]. Here we deal with energy method analysis [12] to solve small deflection problem for thin diaphragm.

In order to do stress analysis for uniformly loaded square thin diaphragm under the effect of pressure load, a two-dimensional trial function [12] can be considered:

$$w = \frac{c}{4}\left[1+\cos\left(\frac{2\pi x}{L}\right)\right]\left[1+\cos\left(\frac{2\pi y}{L}\right)\right]$$

(1)

where w is the trial variational displacement function, c is deflection of the diaphragm at the center (x=0, y=0), L is the edge length of the square diaphragm, and x and y are the in-plane coordinates.

The energy-method analysis [12] with this trial function gives a pressure-deflection equation:

$$P = \left\{c_r\left[\frac{\sigma_o H}{L^2}\right] + c_b\left[\frac{EH^3}{(1-v^2)L^4}\right]\right\}c + c_s f_s(v)\left[\frac{EH}{(1-v)L^4}\right]c^3$$

(2)

where P is the pressure applied, c_r is the coefficient of the residual stress, σ_o is the residual stress, H is the thickness of the membrane, c_b is the coefficient of the diaphragm bending term, c_s is the coefficient of the large-amplitude in plane stretching term, v is the Poisson ratio, and E is Young's modulus of the material.

But, for the present case, membrane is sufficiently thin that the bending term can be neglected (i.e. c_b =0). Further, we assume a small amplitude loading, in which case c_s =0. Thus the equation (2) simplifies to:

$$P = c_r\left(\frac{\sigma_o H}{L^2}\right)c$$

(3)

Since for the case of membrane the stress term should be dominant at small deflections, hence in the equation (3) the right hand side term, linear in c, becomes stress dominated when:

$$\sigma_o \approx \frac{EH^2}{L^2}$$

(4)

Hence, equation (3) becomes:

100

$$P = c_r \left(\frac{EH^3}{L^4} \right) c \tag{5}$$

In order to calculate the total shear stress at the diaphragm edge, we find the x-directed radius of curvature due to bending at the center of the edge (y =0) as follows:

$$\frac{1}{\rho_x} = \left(\frac{\partial^2 w}{\partial x^2} \right)_{x=L/2, y=0} = \left(\frac{2\pi}{L} \right)^2 \frac{c}{2} \tag{6}$$

The magnitude of the x- directed surface stress is

$$\sigma_x = \frac{EH}{2\rho_x} \tag{7}$$

Using equation (6) into equation (7), we find:

$$\sigma_x = \frac{EH}{2} \left(\frac{2\pi}{L} \right)^2 \frac{c}{2} \tag{8}$$

Substituting the value of E from equation (5) into equation (8), σ_x can be expressed as:

$$\sigma_x = \frac{\pi^2}{c_r} \left(\frac{L}{H} \right)^2 P \tag{9}$$

For the membrane limit (assuming $c_b = 0$) [12, 13], $c_r = 13.64$, Thus equation (9) simplifies to:

$$\sigma_x = 0.722 \left(\frac{L}{H} \right)^2 P \tag{10}$$

According to Clark and Wise [14], a full numerical simulation result for the bending of square diaphragm using finite difference methods is:

$$\sigma_x = 0.294 \left(\frac{L}{H} \right)^2 P \tag{11}$$

Our variational solution gives numerical agreement with the exact stress solution to within a factor of 2. The y- directed stress at the center of the edge could be calculated as:

$$\sigma_y = v\sigma_x \tag{12}$$

Referring to Figure 5.1, the two principal-axis stresses, σ_x and σ_y are equivalent to a single net shear stress in the rotated coordinate system given by

$$\sigma = \left(\frac{\sigma_x - \sigma_y}{2} \right) \tag{13}$$

Using (10) and (12) into equation (13), we get:

$$\sigma = 0.722 \left(\frac{L}{H} \right)^2 P \left(\frac{1-v}{2} \right) \tag{14}$$

from equation (14), it clears that thin diaphragm will offer maximum stress and hence maximum sensitivity of the device.

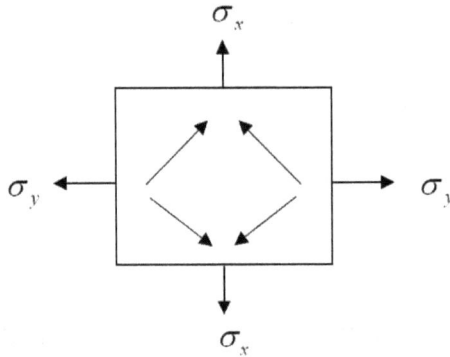

Figure 5.1. The x- and y- directed in plane axial stresses.

Figure 5.2 shows the relation between the maximum allowable pressure applied and the size of square composite membrane made of ($Si_3N_4 +$ $+ SiO_2 + Si_3N_4$) using the membrane thickness as a parameter. Such pattern has also been obtained for a circular diaphragm with diameter of 0.5 mm and the thickness of 5 µm [15].

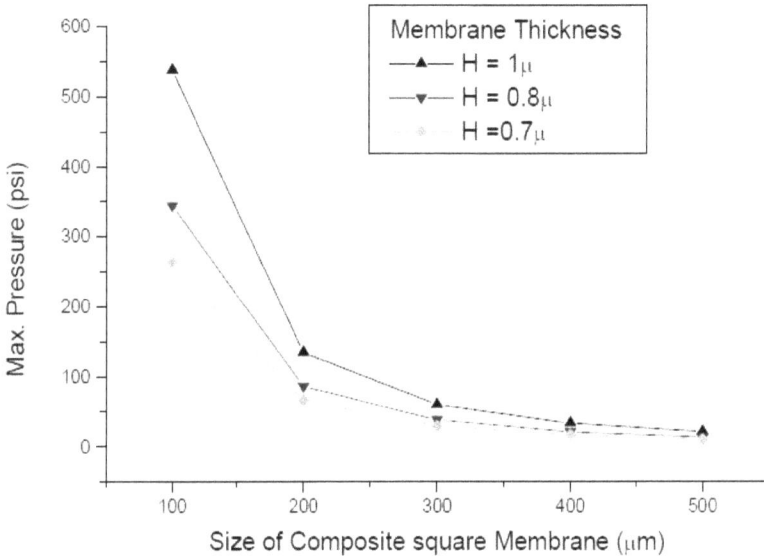

Figure 5.2. Maximum pressures as a function of membrane area for a clamped membrane of composite layer. The membrane thickness is used as a parameter.

5.2. Wheatstone Bridge

The Wheatstone bridge configuration of the resistances over the diaphragm contributes maximum in influencing the sensitivity of the device. This also includes the shape and location of each piezoresistors of the configuration [14]. The ratio of the two resistances in each branch of the Wheatstone bridge is the parameter, which is crucial to optimize the sensitivity of the sensor as a whole [16]. The change in the resistances due to strain on the diaphragm develops potential difference, which translates the applied pressure. In order to increase the sensitivity of the bridge, the resistances are placed on the diaphragm so as to get maximum change in potential difference.

In order to understand the basics of the placement of the resistors on the diaphragm, consider the four resistances of the bridge R_1, R_2, R_3 and R_4, as shown in Figure 5.3.

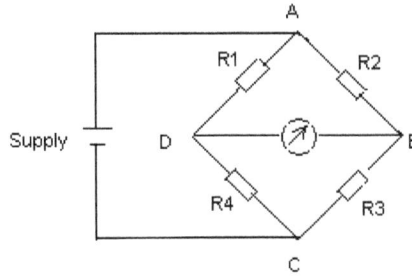

Figure 5.3. Wheatstone bridge configuration.

The bridge is biased between the points A and C and the potential difference developed across the points D and B is monitored. In case of ideal situation of point D and B at the same potential level, following relation holds:

$$\frac{R1}{R2} = \frac{R4}{R3}$$

However, for the purpose of sensing applications, a potential difference between points D and B is required, which is possible under following condition;

$$\frac{R1}{R2} \neq \frac{R4}{R3}$$

The potential developed between points D and B; V_{out} is a measure of pressure sensor sensitivity, and is given by

$$V_{out} = V_{in}\left[\frac{R_1}{R_1 + R_2} - \frac{R_4}{R_4 + R_3}\right]$$

Obviously, for ideal condition (under no pressure), there is no output voltage.

The magnitude of V_{out} can be adjusted with the help of the two ratios of the piezoresistances as shown above. In order to achieve highest possible sensitivity following condition is required to meet;

$$\left| \frac{R1}{R2} - \frac{R4}{R3} \right| = \text{Maximum}$$

It can be deduced from the above that the most desirable condition is to allow R_1, R_3 and R_2, R_4 to vary in the opposite way. This can be accomplished on a strained diaphragm where stress profile is composed of tensile and compressive fringes. In this case the piezoresistors change their values in contrast while interacting with tensile and compressive regions of the stress. This is possible only when two of the resistors of the bridge are placed in the region of one type of stress while remaining two are placed in the region of another type of stress. But for very small diaphragm, as in our case, to accommodate all the four resistors on the diaphragm is difficult. In such case, one way to obtain the above condition of maximum sensitivity is to place two of the resistors, either R_1 and R_3 or R_2 and R_4 on the diaphragm where stress will occur and the other two outside the diaphragm where stress will not occur. In the design presented R_2 and R_4 are placed on the diaphragm and R_1 and R_3 outside the diaphragm. Figure 5.4 shows the top view (Layout) of the device.

If R_2 and R_4 are on the diaphragm (called as effective resistance R_e) and R_1 and R_3 outside the diaphragm (called as non-effective resistance R_n) then under the pressure applied on the diaphragm the output voltage signal is given by:

$$V_{out} = \frac{2 \cdot R_n \cdot \Delta R_e}{\left(R_e + R_n \right)^2} V_{in}$$

,

where ΔR_e is the change in effective resistance.

In Figure 5.4, the hatch-lined objects are the resistors, square box with dotted lines is the diaphragm, the two blacken rectangles are the etch holes through which etch material (wet etchant) enters in the bulk of the substrate, five square boxes towards the periphery are the contact pads and the lines from resistors to these pads are the metal contact lines. The corresponding design parameters are mentioned below:

Chip size: 1 mm × 1 mm
Diaphragm size: 100 μm × 100 μm
Diaphragm thickness: 0.7 μm
Resistor's configuration: half- Wheatstone bridge
Resistor's length: 110 μm
Resistor's width: 10 μm
Resistor's pad: 20 μm × 20 μm
Resistor's thickness: 1.0 μm
Metal contact pad: 100 μm × 100 μm
Metal contact line width: 20 μm
Resistor's value: 0.22 kΩ
Sheet resistivity of resistors: 20 Ω /

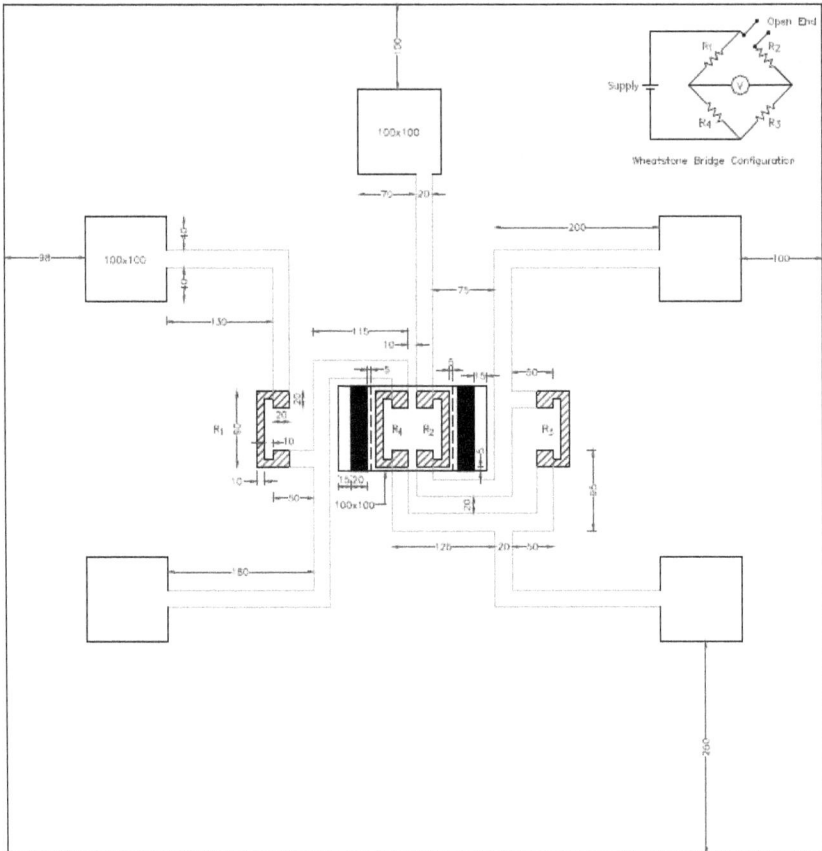

Figure 5.4. Schematic diagram of the top view (Layout)
of the micro pressure sensor.

The simulated stress profile, shown in Figure 5.5, is the result of analysis of the device membrane using ANSYS software. From this, it is clear that stress is maximal at the edges and decreases as one moves away towards the center. Accordingly, the design layout with the resistors oriented in the transverse direction should give more sensitivity, as the resistors will experience maximum stress and hence maximum piezo-effect. The shapes of the resistors are chosen to ensure maximum length that can be accommodated in the diaphragm. The actual size of the diaphragm is kept 100 μm × 100 μm to realize a miniature sized pressure sensor- a micro sensor so that it will be compatible with other microelectronic devices. Keeping the width of the resistors 10 μm and a square of 20 μm × 20 μm for contacts of contact lines, the total length comes out to be 150 μm. Since the square of 20 μm × 20 μm is on either ends and is to be connected with the conducting metals, the net length of the resistors is only 110 μm for 100 μm × 100 μm diaphragm.

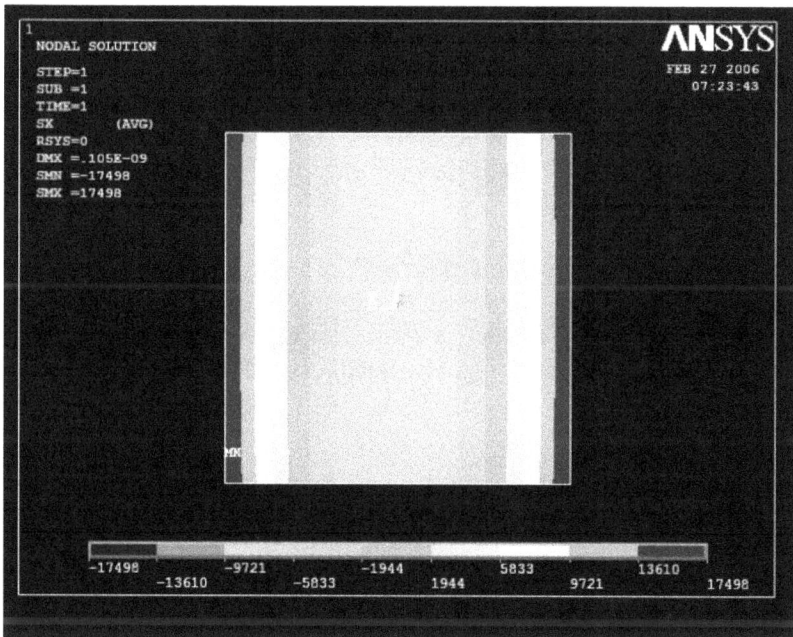

Figure 5.5. Simulated stress profile of a strained composite membrane of thickness 0.8 μm on application of 344-psi pressure.

5.3. Piezoresistors

Piezoresistors deposited over the diaphragm are made of boron-doped polysilicon because of good piezoeffect. Polysilicon has better stability and can be used in operating temperature up to 200 °C [17]. Certain materials like Si or poly Si are sensitive to change their resistance resulting from stress applied to the crystal lattice. Resistance, in

particular, is dependent on the changes in length caused by stress. These Resistive changes are not isotropic, and can be divided into two independent functions, one component parallel to the direction of stress, and other component perpendicular to it, in the form of following expression:

$$\frac{\Delta R}{R} = \pi_l \sigma_l + \pi_t \sigma_t$$
,

where π_l and π_t are the piezoresistive coefficients in longitudinal and transverse directions, respectively. Stresses in longitudinal and transverse directions are designated by σ_l and σ_t . These coefficients depend on the orientation of resistors and as well as are the functions of temperature and doping concentration. The values of piezoresistive coefficients decrease with increase in impurity concentration and increase in temperature.

Low value of the resistors is preferred because when there is a slight change in the resistor's value due to piezoresistive effect, the change in resistance compared to the original resistor's value will be more noticeable rather than with higher valued resistors.

Since by relation,

$$\text{Resistance, } R = (\rho \times l)/A,$$

where l is the Length,
$A = t \times w$
(Here t is the thickness, and w is the width).

Hence, the formula can be rewritten as:

$$\text{Resistance, } R = (\rho/t) \times (l/w)$$

or

$$R = (\rho_s \times l)/w,$$

where $(\rho_s = \rho/t)$ is called sheet resistivity of the surface of the resistor. The value of sheet resistivity can be controlled by:

1) Polysilicon layer thickness.

2) Type of dopant and the temperature at which dopant is diffused (as shown in Figure 3.2 in Chapter 3).

With increasing polysilicon layer thickness the sheet resistivity decreases. For 1.0 μm thick poly Si layer the sheet resistivity can be brought down in the range of 15-20 Ω/. This way by controlling the polysilicon layer thickness, values of the polysilicon resistors can be designed for the given diaphragm.

The dimension of the sacrificial layer (polysilicon), to realize cavity later on, is kept 180 μm × 100 μm with 40 μm each extending on either sides from the diaphragm dimensions. This is done so that etch holes of sufficient dimension can be accommodated beyond the actual diaphragm but within the dimension of sacrificial layer. This condition is necessary for sealing the cavity in the later stage. The dimension of the etch hole is kept 20 μm × 100 μm, maintaining 5 μm from the diaphragm edge on either side. This dimension for etch hole is sufficient to provide passage for wet etchant to create cavity under the diaphragm. Also the etch holes should not be in contact with the diaphragm edges otherwise the diaphragm will break down.

The width of the metal contact lines are kept 20 μm and the minimum gap between two contact lines is tried to keep at least twice its width. The dimension for metal contact pads is kept 100 μm × 100 μm. Here five pads are used instead of four since one of the arms is opened. This provision is for the purpose of balancing the bridge externally if required and if not required then by shorting, it can be used normally. Leaving sufficient space on all the sides, the size of a single chip comes up to be 1 mm × 1 mm. This is a pretty small size. Leaving grid size of 0.1 mm (the size that can be comfortably run by die-slicer blade), the number of devices that can be fabricated on a single wafer of 2 inches diameter after leaving sufficient space for alignment marks on either side are an array of 33 × 33 giving 1089 devices. Keeping all these in views, the layouts for the device are designed using the software L-Edit (layout editor) for layout design.

References

[1]. E. Suhir, Structural analysis in microelectronic and fiber optic systems, Vol. I, *Van Nostrand Reinhold*, 1991.

[2]. J. A. Voorthuyzen, and P. Bergveld, The influence of tensile forces on the deflection of circular diaphragms in pressure sensors, *Sensors and Actuators*, 6, 1984, pp. 201-213.

[3]. R. Schellin, G. Hess, W. Kuhnel, C. Thielemann, D. Trost, J. Wacker, and R. Steinmann, Measurements of the mechanical behavior of micromachined silicon and silicon-nitride membranes for microphones, pressure sensors, and gas flow meters, *Sensors and Actuators*, A, 41-42, 1994, pp. 287-292.

[4]. D. Maier-Schneider, J. Maibach, and E. Obermeier, A new analytical solution for the load-deflection of square membranes, *Journal of Microelectromechanical Systems*, 4, 4, Dec., 1995, pp. 238-241.

[5]. H. E. Elgamel, Closed form expressions for the relationships between stress, diaphragm deflection, and resistance change with pressure in silicon piezoresistive pressure sensors, *Sensors and Actuators A*, 50, 1995, pp. 17-22.

[6]. R. Steinmann, H. Friemann, C. Prescher, R. Schellin, Mechanical behavior of micromachined sensor membranes under uniform external pressure affected by in-plane stresses using a Ritz method and Hermite polynomials, *Sensors and Actuators A*, 48, 1995, pp. 37-46.

[7]. S. Timoshenko and S. Woinosky-Krieger, Theory of Plates and Shells, *McGraw Hill Classic Textbook Reissue*, 1987.

[8]. H. M. Berger, A new approach to the analysis of large defilections of plates, *Journal of Applied Mechanics*, Vol. 22, 1955, p. 465-472.

[9]. R. Szilard, Theory and Analysis of Plates: Classical and numerical Methods, Englewood Cliffs, *Prentice-Hall*, 1974.

[10]. C.-Y. Chia, Nonlinear Analysis of Plates, *McGraw-Hill*, New York, 1980.

[11]. S. P. Timoshenko and S. Woinowsky-Krieger, Theory of Plates and Shells, 2nd Ed., *McGraw-Hill*, New York, 1970.

[12]. S. D. Senturia, Microsystem Design, *Kluwer Academic Publishers*, USA, 2001.

[13]. P. Lin and S. D. Senturia, The in-situ measurement of biaxial modulus are residual stress of multi-layer polymeric thin films, in Thin Films: Stress and Mechanical Properties II, M. F. Doerner, W. Oliver, G. M. Pharr, and F. R. Brotzen (eds.), *Material Research Society Symposium Proceedings*, San Francisco, CA, Vol. 188, April 16-19, 1990, pp. 41-46.

[14]. S. K. Clark and K. D. Wise, Pressure sensitivity in anisotropically etched thin diaphragm pressure sensors, *IEEE Transaction on Electron Devices*, Vol. ED- 26, December, 1979, pp. 1887-1896.

[15]. Samaun, K. D. Wise, and J. B. Angell, An IC piezoresistive pressure sensor for biomedical instrumentation, *IEEE Transaction on Biomedical Engineering,* Vol. BME-20, No. 2, March, 1973, pp. 101-109.

[16]. J. Akhtar, B. B. Dixit, B. D. Pant, V. P. Deshwal, B. C. Joshi, Ram Gopal, R. N. Soni, Ravi Bhatia, A. K. Bagchi, A. K. Gupta, Satish Kumar, Om Prakash, and Mahesh Kumar, Design and development of polysilicon piezoresistive pressure sensor based on MEMS technology, Technical Report, *CEERI*, Pilani, 2002.

[17]. R. Timothy Edwards, Microfabrication Lab Project Report: A Simple CMOS Pressure Sensor, December, 1994.

Chapter 6

Fabrication of Micro Pressure Sensor (using Fron-side Lateral Etching Technology)

6.1. Front-Side-Etching Technology

Front side etching technique is the most advanced and recently developed MEMS technology. It takes the advantages of both the bulk and surface micromachining technologies (The details of both technologies have been discussed in Chapter 1).

Continuous advancement in the micro-fabrication technology has been emerged into more compatibility with the microelectronics processes systematically leading towards the realization of system on a chip concept. Starting from bulk micromachining to surface micromachining, new etching techniques are being developed in order to fabricate micrometer size electro-mechanical devices with controlled physical dimensions on the device grade crystalline silicon. Whereas the bulk micromachining of (100) silicon has been used for silicon-based diaphragm [1], a thin membrane type diaphragm has been formed using surface micromachining techniques involving external layers of compatible materials [2]. There are processes in the literature where either bulk micromachining or surface micromachining has been used separately for different types of micromechanical device fabrication.

Bulk micro-machined pressure sensors are one of the earliest products made by silicon micro-machining technology [3]. Today, many companies fabricate and sell bulk micro-machined pressure sensors for automobile, industrial, and bio-medical applications. The measurement range for these pressure sensors can be as high as 10,000 Psi with excellent reliability. On the other hand, pressure sensors based on surface micro-machined diaphragms were first suggested and fabricated in the 1980s [4]. Thin film deposition and reactive sealing technologies were used to fabricate polysilicon diaphragms with cavities underneath. The back side silicon wet etching process which has been used for bulk micro-machined pressure sensors was avoided [5]. These surface micro-machined pressure sensors may be more attractive than the bulk micro-machined ones because of the following reasons. First, Bulk micro-machined pressure sensors require anisotropic silicon etching [6] to create thin diaphragms from the back side of silicon wafers. This process consumes large areas [7]. For example, if a standard four inch wafer with thickness of 500 µm is used, an area of about 800 × 800 µm^2 is required to make 100 × 100 µm^2 diaphragm. However, an area of only 100 × 100 µm^2 is needed to make a surface-micro-machined diaphragm. Second, bulk-micro-machined pressure sensors require post processing including glass to silicon bonding before the final packaging process. Surface micro-machined pressure sensors are ready for packaging after the micro-machining processes. Finally, surface micro-machining is easier to be integrated with IC processes for additional signal processing or device functionality.

The term front-side etching has been emerged due to bulk micromachining, which is usually performed on the backside of silicon wafer, making thus, back to front alignment, a prime process-limiting factor. Owing to back to front alignment requirements, bulk micromachining loses its compatibility with conventional microelectronic processes with additional factor of non-planar silicon surfaces. In order to eliminate back to front alignment and avoiding non-planarity, surface micromachining has been developed with inherent drawbacks related to diaphragm sensitivity and application oriented limitations [8, 9]. Front-side etching processes take care for the advantages of the bulk micro-machining and also provide a process platform similar to surface micromachining with enhanced compatibility to common microelectronic processes. Front-side-etching technology can be classified as follows.

6.1.1. Front Side Normal Etching Technology

This technique is based on etching via hole from the front side of the silicon wafer. In case of via hole technology, concave window for anisotropic etching becomes effective for controlled under etching, resulting into a thin membrane on the edges of the membrane area. Presence of holes on the membrane further limits the applications of the proposed techniques. This technique focuses on the development of the silicon membrane fabrication. Fabrication of thin membrane has been an important aspect in common microelectronic device owing to its numerous industrial applications. A flat membrane of silicon nitride and silicon dioxide has been fabricated using a process based on bulk micromachining [10]. A thin silicon membrane has been reported using front side normal etching technique [11, 12]. The entire micro-machining process steps can be accomplished at the front side of the silicon substrate, due to the etching via whole technology being applied to form the silicon membrane. The fabrication processes comprise several steps, as shown in Figure 6.1 (a)-(e). Figure 6.1 (a) illustrates a P-type (100) silicon substrate with an N epi-layer and thermal oxide. In order to form the etching holes, the silicon wafer was etched by deep reactive ion etching (DRIE), as shown in Figure 6.1 (b). The top view photograph of the etching holes is shown in Figure 6.1 (c). Since the electrochemical etch-stop technique was adopted, the silicon wafer was immersed in the etching solution such as KOH. The etchant will be through the etching holes and in contact with the P- type substrate. The P- type substrate was etched anisotropically. Because the etching rate highly depends on the silicon crystal orientation, the etchant will etch any <100> silicon until a pyramidal pit is formed. Consequently, the P- type substrate under the N-epi layer will form little V-grooves in the initial reactions. While the little V-grooves touch each other, the convex corners are being undercut by the etchant and as a result the larger V- grooves are formed. The V-grooves are illustrated as dash lines in Figure 6.1 (d). Even though the shapes of the etching holes are circular, the eventual geometry of the dash line areas will be squares as seen from the top view. Finally, a silicon membrane with etching holes was fabricated as shown in Figure 6.1 (e). For gauge measurement, the etching holes were sealed by a high viscosity and photosensitive polymer such as polyimide.

From the above fabrication processes it can be observed that the etching via whole technology can accomplish a front side etching process to replace the back-side etching process on the sensor fabrication. By employing the front side normal etching process, the chip size can be reduced by at least 50 % over the conventional method.

6.1.2. Front Side Lateral Etching Technology

In this technique, the silicon substrate under the defined membrane area is etched away in a controlled manner so that a workable membrane is realized. This is accomplished through channels on the side of the membrane area.

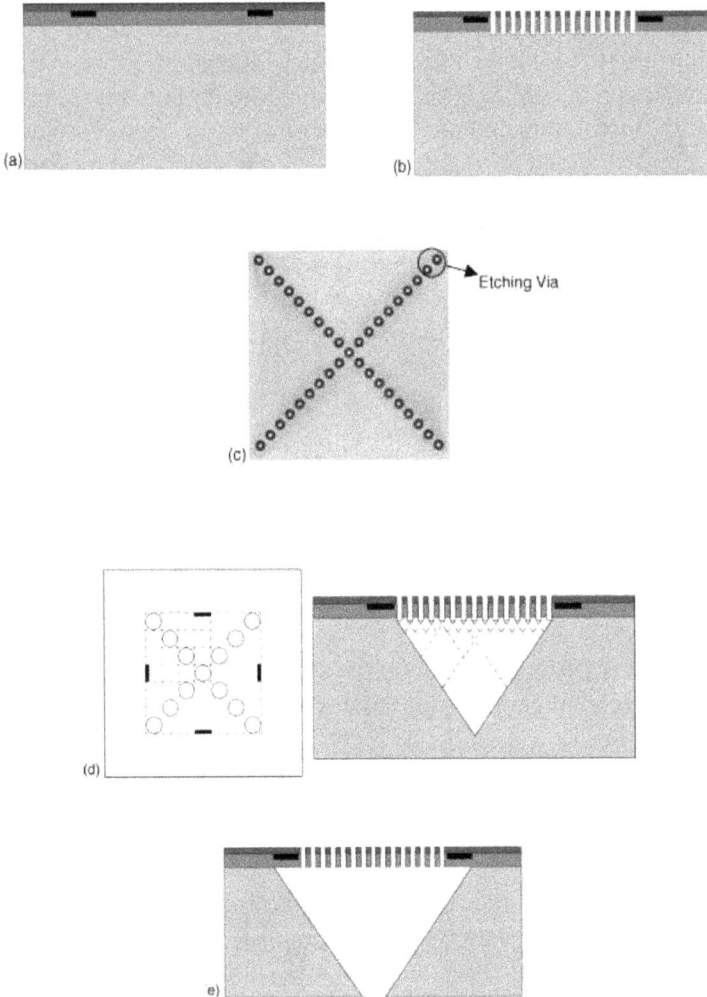

Figure 6.1. Fabrication process flow of membrane using front side normal etching technology.

After removing the sacrificial layer from the identified channel region, anisotropic etching of the silicon substrate is done in to aqueous KOH at mild temperature. The square geometry of the silicon window results into a conical cavity. The etching mechanism of silicon through side channels, as shown in Figure 6.2, takes place in successive V-groove with increasing size under the both sides of the membrane. The lateral etching through channels on the sides of the membrane, slowly etches towards the center of the membrane so that one wall of (111) silicon is fixed and another wall is dynamic owing to more polysilicon etch on the top. The dashed lines in the Figure 6.2 indicate the moving etched wall in the direction of arrow. The final shape of the etched cavity looks like as shown in Figure 6.3. The hanging membrane is stayed on the other two sides with fixed edges. The membrane edge towards the channels remains free introducing more sensitivity.

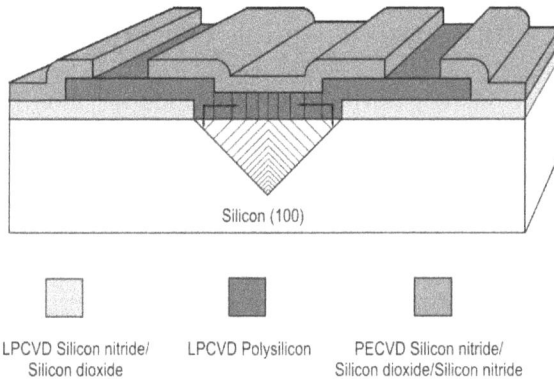

▢	▪	▢
LPCVD Silicon nitride/ Silicon dioxide	LPCVD Polysilicon	PECVD Silicon nitride/ Silicon dioxide/Silicon nitride

Figure 6.2. Schematic 3-D details of the membrane and etching mechanism under the membrane.

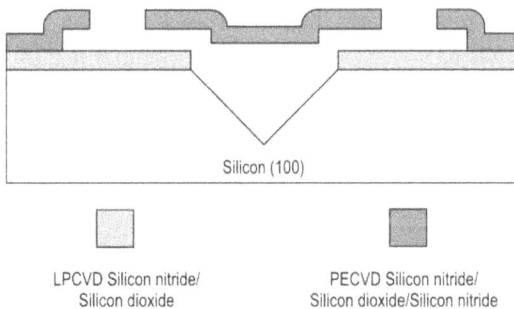

▢	▢
LPCVD Silicon nitride/ Silicon dioxide	PECVD Silicon nitride/ Silicon dioxide/Silicon nitride

Figure 6.3. Schematic diagram of the membrane and etched cavity.

The lateral etching rate of the final V-groove formation has been found seven times slower than normal etching of the square window of the same size with rest of similar parameters. The lateral front-side etching, therefore, has been found significantly different from front-side normal etching structures [11].

In the present work, we use front-side lateral etching technology to fabricate membrane type micro pressure sensor of die size 1 mm × 1 mm – Absolute micro pressure sensor. This technology is capable to control micrometer feature size machining in addition to being compatible with conventional microelectronics. By using front-side lateral etching technique we create a conical cavity under the membrane in the bulk of substrate through channels on the side of the membrane area as shown in Figure 6.3, which is sealed later on by LPCVD or APCVD techniques.

With front-side etching technique, deposition of thin films on the surface of the wafers and removing portions of one or more layers to realize desired mechanical structures are done plus 3-dimensional features are etched using anisotropic etching into the areas of Si-substrate to form cavity and diaphragm. Here the etched out portions are in few tens of microns so no large wastage as in the case of bulk. Also since cavity is formed underneath the membrane there is no chance of adhesion problems as in the case of surface micro-machined structure. This technology thus enables us to realize miniature size rugged structures that can withstand vibrations, shocks and can easily integrate into IC processes.

From the technology point of view, development of a compatible process to integrate the MEMS based processes with conventional microelectronics is the main issue for the realization of these pressure sensors. In the MEMS technology, anisotropic etching of <100> silicon substrate is carried out for various purposes such as formation of diaphragm and similar other processes.

The fabrication detail of membrane type micro pressure sensor using front side lateral etching technique is given in the following section.

6.2. Fabrication Process Sequence Optimization

A p-type silicon wafer of diameter 2 inch with (100) orientation and thickness of 280±20 microns is the starting substrate, as shown in

Figure 6.4. For a proper V-shape cavity formation, silicon with (100) is the ultimate choice. A sequence of processes to realize the device is decided depending upon the resources available in the laboratory. Depending upon the outlined processes, masks are designed and fabrication is preceded. Sequence of the process is important and needs some runs leading to modifications in the sequence prior to reach its optimization.

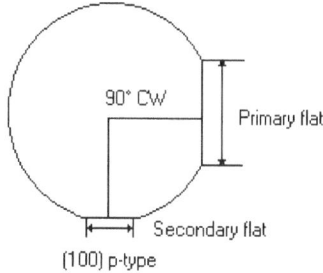

Figure 6.4. A p-type Si wafer with (100) orientation.

The fabrication can be accomplished by two processes. One process in which cavity is formed at final stage and in other cavity is formed at initial stage. Both of the process sequences are described as follows.

6.2.1. Fabrication Process Sequence (I)

In this process the cavity is formed at final stage. The process starts with cleaning the wafer followed by the following sequence:

1. SiO_2 thermally grown.

2. Si_3N_4 (LPCVD).

3. PLG-1 (window opening) + RIE of Si_3N_4 + Wet etching of SiO_2.

4. Polysilicon deposition (sacrificial layer) by LPCVD.

5. PLG-2 + RIE (for poly etch).

6. PECVD of $(Si_3N_4 + SiO_2 + Si_3N_4)$, then annealing.

7. LPCVD polysilicon.

8. Boron doping of Poly Si for piezoresistors.

9. PLG-3 + Poly etch (for defining resistors).

10. Metallization of Al for resistor pads and contact lines.

11. PLG-4 + Wet etching of Al.

12. APCVD SiO_2 + PECVD Si_3N_4.

13. PLG-5 + RIE of composite layer (Si_3N_4 + SiO_2 + Si_3N_4).

14. KOH Etching for Poly Si & Si in bulk of substrate (formation of cavity)

15. APCVD SiO_2 + PECVD Si_3N_4 (For sealing the cavity and passivation).

16. PLG-6 + RIE (SiO_2 + Si_3N_4).

17. PLG-7 + RIE (for grid).

The corresponding fabricated final device i.e. the absolute micro pressure sensor (3-D view) by using process sequence (I) is shown below.

3-D view of Device

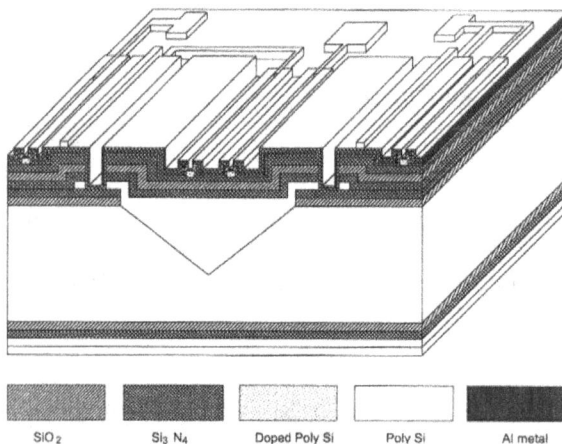

SiO₂   Si₃N₄   Doped Poly Si   Poly Si   Al metal

The corresponding flow chart for process sequence (I) is given below.

Fabrication process sequence (I)

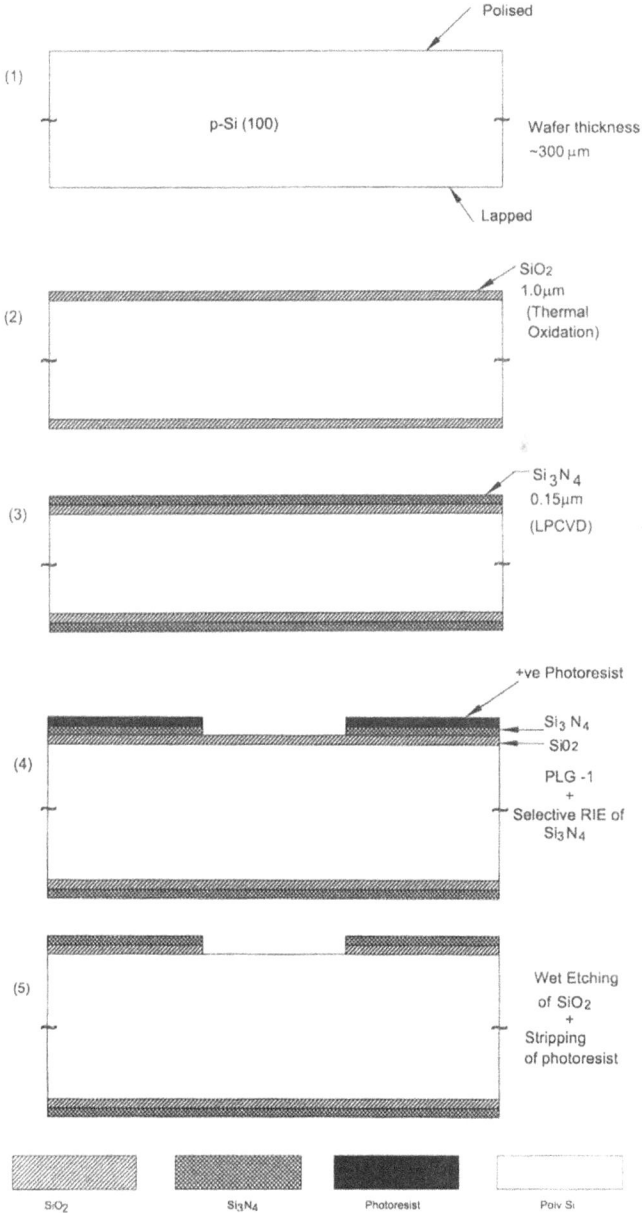

(1) p-Si (100) — Polised — Wafer thickness ~300 μm — Lapped

(2) SiO_2 1.0μm (Thermal Oxidation)

(3) Si_3N_4 0.15μm (LPCVD)

(4) +ve Photoresist — Si_3N_4 — SiO_2 — PLG -1 + Selective RIE of Si_3N_4

(5) Wet Etching of SiO_2 + Stripping of photoresist

SiO_2 Si_3N_4 Photoresist   Poly Si

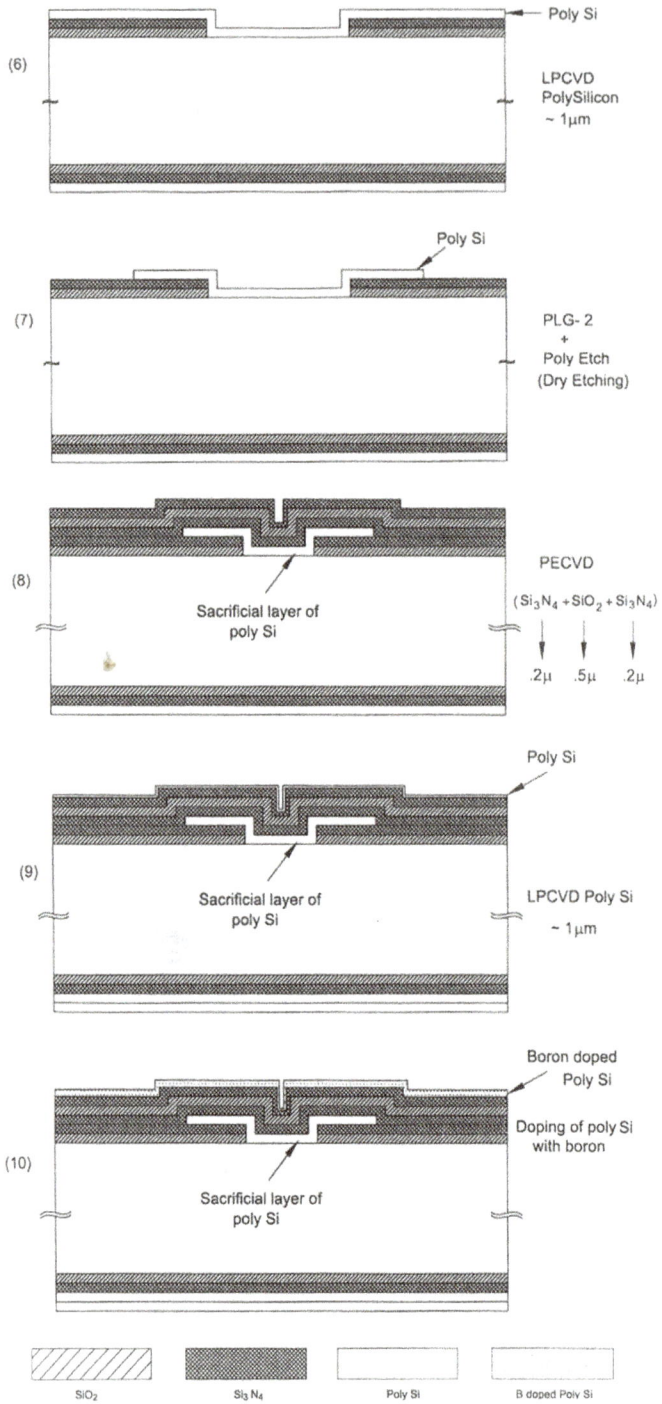

(6)   Poly Si

LPCVD
PolySilicon
~ 1µm

(7)   Poly Si

PLG- 2
+
Poly Etch
(Dry Etching)

(8)   Sacrificial layer of poly Si

PECVD

$(Si_3N_4 + SiO_2 + Si_3N_4)$

.2µ   .5µ   .2µ

(9)   Poly Si

Sacrificial layer of poly Si

LPCVD Poly Si
~ 1 µm

(10)   Boron doped Poly Si

Doping of poly Si with boron

Sacrificial layer of poly Si

| SiO_2 | Si_3N_4 | Poly Si | B doped Poly Si |

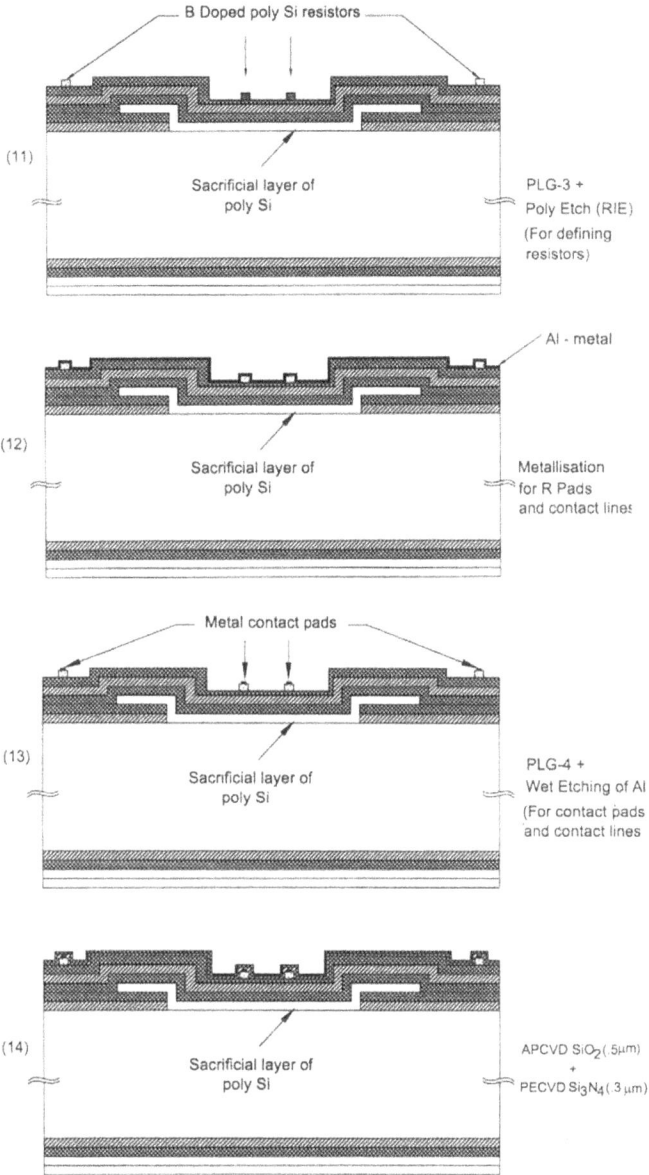

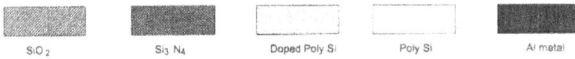

123

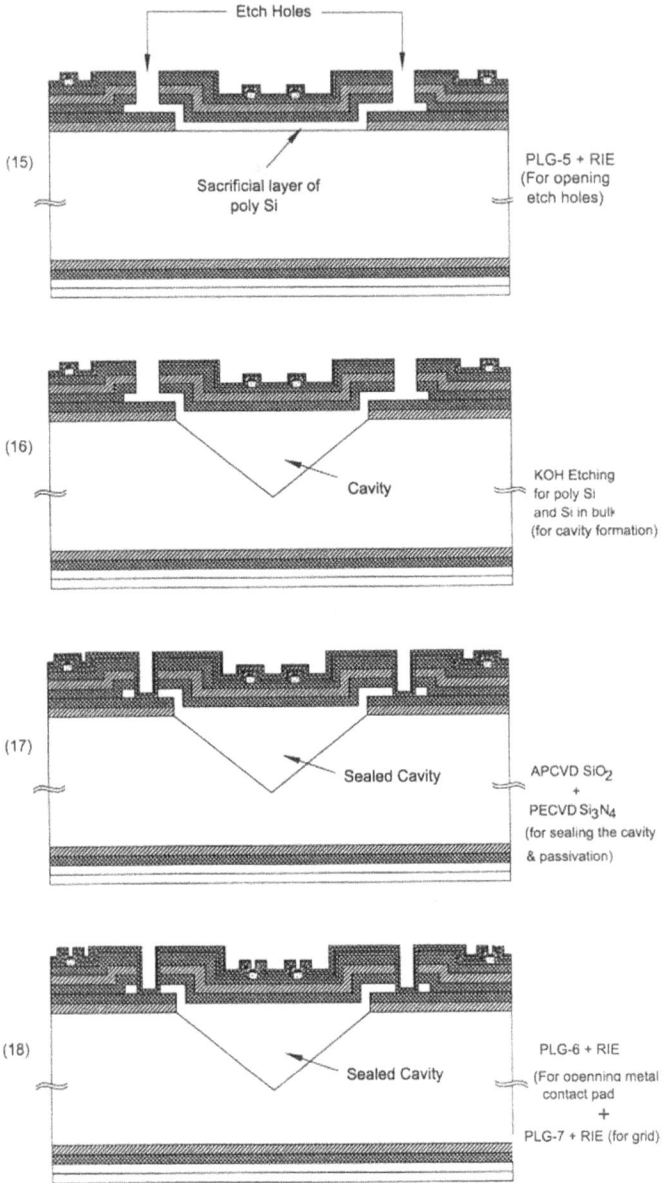

(15) PLG-5 + RIE (For opening etch holes)

Etch Holes

Sacrificial layer of poly Si

(16) KOH Etching for poly Si and Si in bulk (for cavity formation)

Cavity

(17) APCVD SiO$_2$ + PECVD Si$_3$N$_4$ (for sealing the cavity & passivation)

Sealed Cavity

(18) PLG-6 + RIE (For opening metal contact pad + PLG-7 + RIE (for grid)

Sealed Cavity

SiO$_2$ Si$_3$N$_4$ Doped Poly Si   Poly Si   Al metal

3-D view of Device

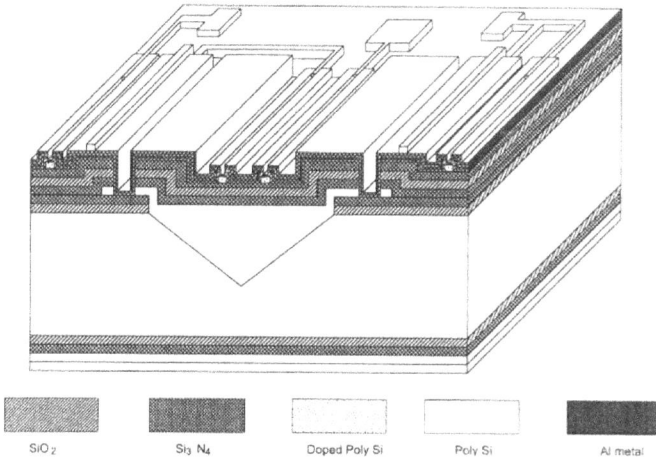

| SiO$_2$ | Si$_3$ N$_4$ | Doped Poly Si | Poly Si | Al metal |

6.2.2. Fabrication Process Sequence (II)

In this process the cavity is formed at initial stage. The process starts with cleaning the wafer followed by the following sequence:

1. SiO$_2$ thermally grown.

2. Si$_3$N$_4$ (LPCVD).

3. PLG-1 (window opening) + RIE of Si$_3$N$_4$ + Wet etching of SiO$_2$.

4. Polysilicon deposition (sacrificial layer) by LPCVD.

5. PLG-2 + RIE (for poly etch).

6. LPCVD (Si$_3$N$_4$ + SiO$_2$ + Si$_3$N$_4$).

7. PLG-3 + RIE (Si$_3$N$_4$ + SiO$_2$ + Si$_3$N$_4$).

8. KOH Etching for Poly Si & Si in bulk of substrate (for cavity formation).

9. LPCVD Poly Si (for sealing of cavity).

125

10. Doping of polysilicon with boron (formation of resistors).

11. PLG-4 + Poly etch (RIE).

12. Metallization of Al for resistor's pads & contact lines.

13. PLG-5 + wet etching of Al.

14. PECVD Si_3N_4 (for passivation).

15. PLG-6 + RIE of Si_3N_4 (for opening contact pad).

16. PLG-7 + RIE (for grid).

The corresponding fabricated final device i.e. the absolute micro pressure sensor (3-D view) by using process sequence (II) is shown below:

3-D view of Device

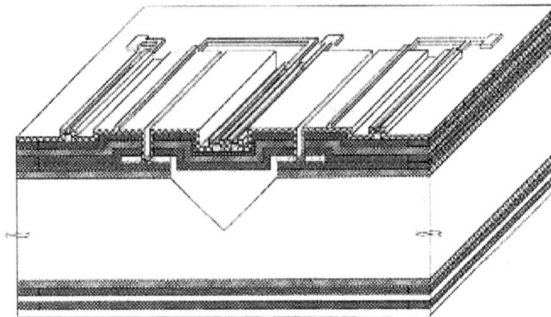

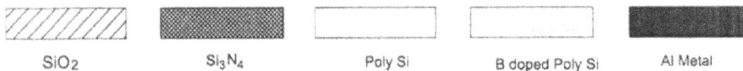

| SiO2 | Si3N4 | Poly Si | B doped Poly Si | Al Metal |

The corresponding flow chart for process sequence (II) is given below:

Fabrication Process Sequence (II)

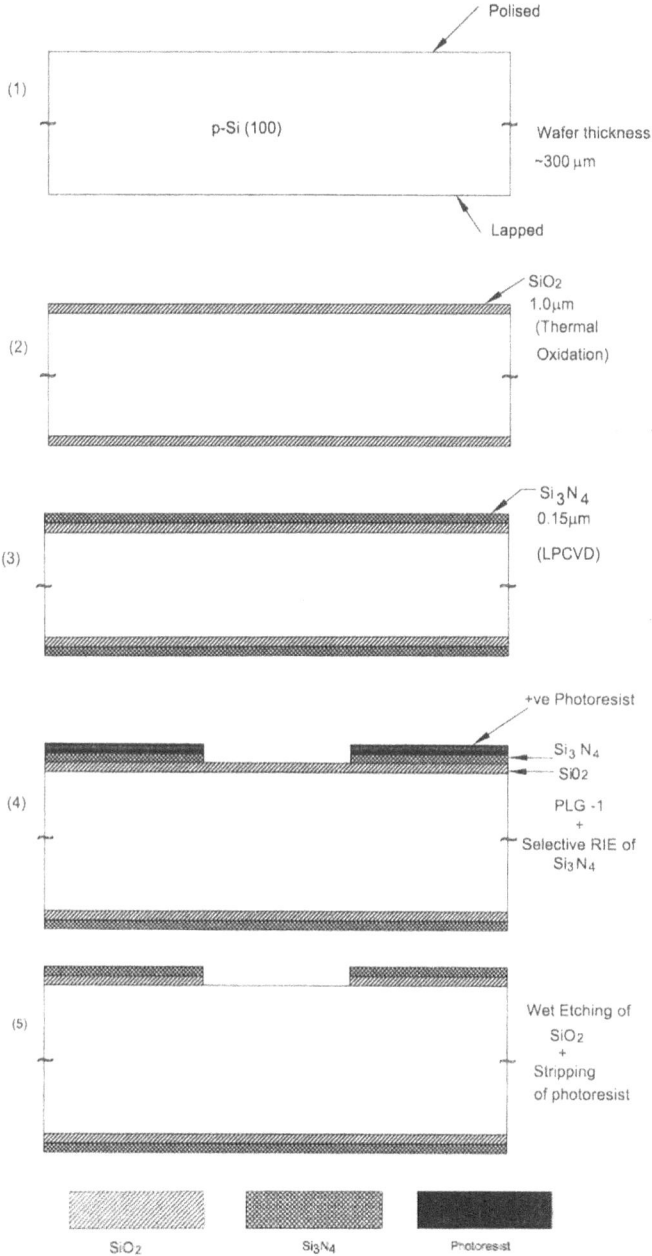

(1) Polised

p-Si (100)

Wafer thickness
~300 μm

Lapped

(2) SiO_2
1.0μm
(Thermal
Oxidation)

(3) Si_3N_4
0.15μm
(LPCVD)

(4) +ve Photoresist
Si_3N_4
SiO_2
PLG -1
+
Selective RIE of
Si_3N_4

(5) Wet Etching of
SiO_2
+
Stripping
of photoresist

SiO_2 Si_3N_4 Photoresist

127

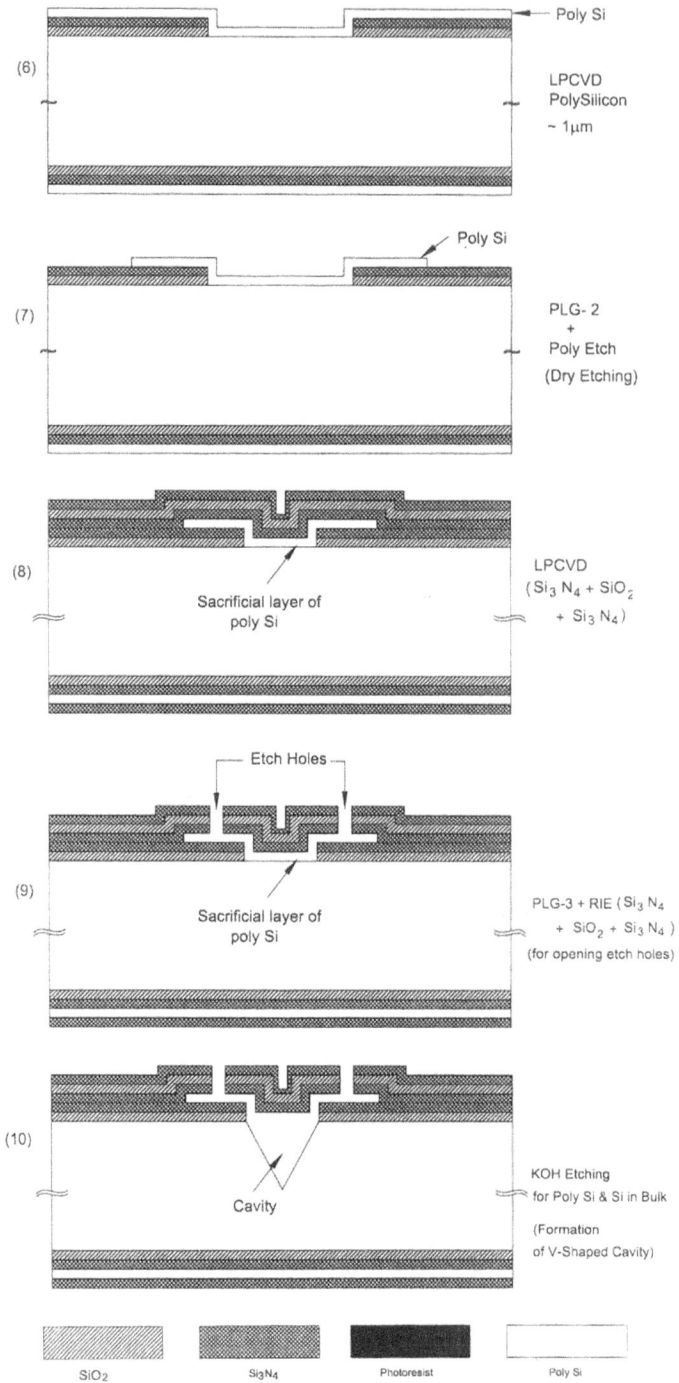

(6)   Poly Si

LPCVD
PolySilicon
~ 1µm

(7)   Poly Si

PLG- 2
+
Poly Etch
(Dry Etching)

(8)

Sacrificial layer of
poly Si

LPCVD
($Si_3 N_4$ + SiO_2
+ $Si_3 N_4$)

(9)

Etch Holes

Sacrificial layer of
poly Si

PLG-3 + RIE ($Si_3 N_4$
+ SiO_2 + $Si_3 N_4$)
(for opening etch holes)

(10)

Cavity

KOH Etching
for Poly Si & Si in Bulk

(Formation
of V-Shaped Cavity)

| SiO_2 | Si_3N_4 | Photoresist | Poly Si |

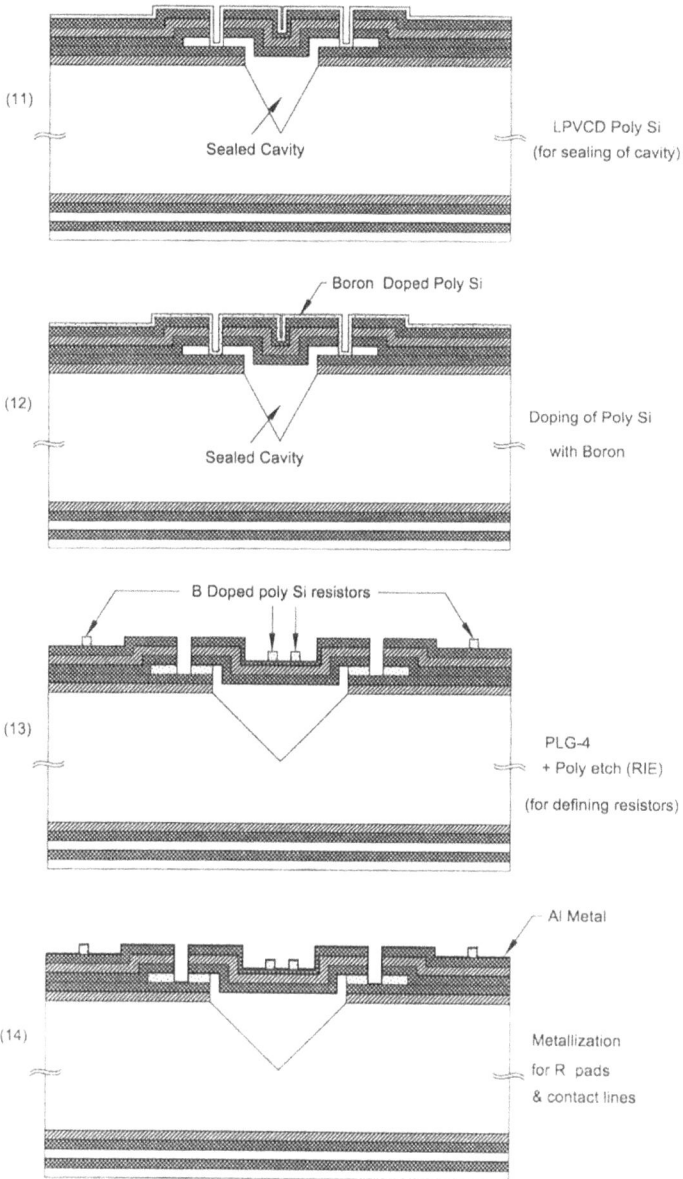

(11)

Sealed Cavity

LPVCD Poly Si
(for sealing of cavity)

Boron Doped Poly Si

(12)

Sealed Cavity

Doping of Poly Si
with Boron

B Doped poly Si resistors

(13)

PLG-4
+ Poly etch (RIE)
(for defining resistors)

Al Metal

(14)

Metallization
for R pads
& contact lines

| SiO₂ | Si₃N₄ | Poly Si | B doped Poly Si |

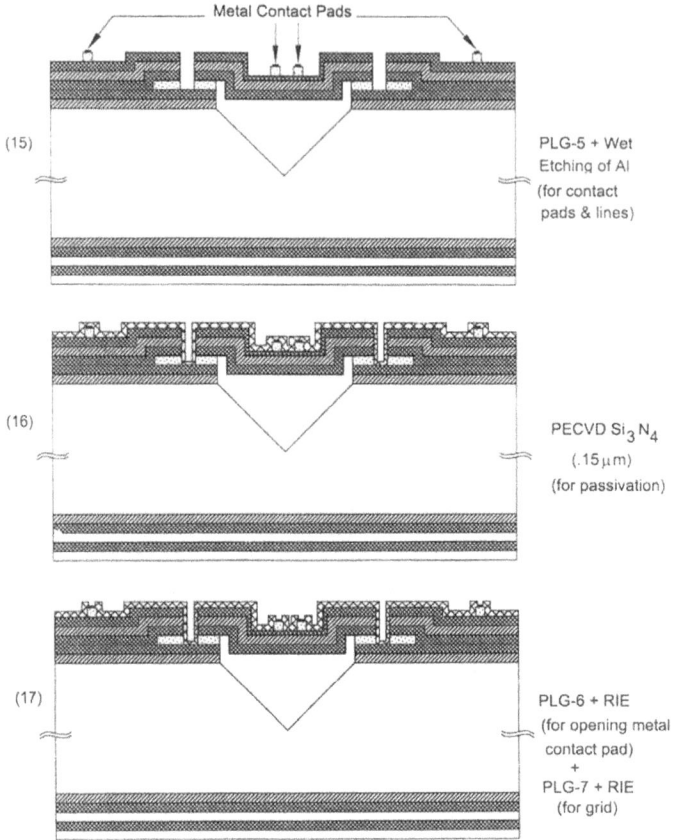

Metal Contact Pads

(15)   PLG-5 + Wet
Etching of Al
(for contact
pads & lines)

(16)   PECVD Si_3N_4
$(.15\mu m)$
(for passivation)

(17)   PLG-6 + RIE
(for opening metal
contact pad)
+
PLG-7 + RIE
(for grid)

3-D View of the Device

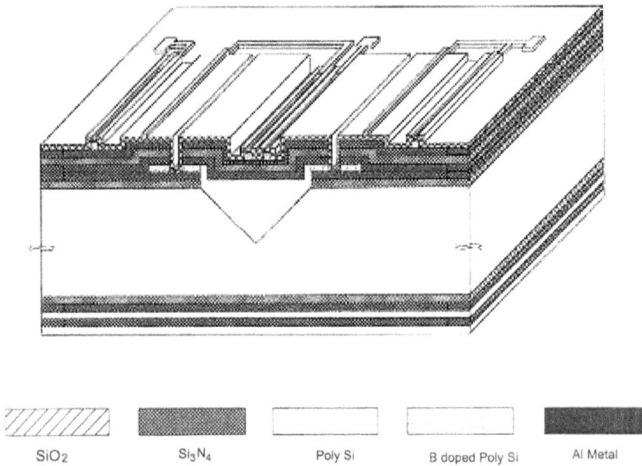

SiO_2	Si_3N_4	Poly Si	B doped Poly Si	Al Metal

130

6.3. Fabrication Process Detail

6.3.1. Detail of Process Sequence (I)

The process starts with cleaning the wafer. The wafer is cleaned in the following steps:

(a) Degreasing

(b) RCA-1

(c) RCA-2

Then wafer is rinsed properly in DI water and dried on hot plate. SiO_2 is thermally grown as thermal growth exhibits better adhesion of SiO_2 with the underlying Si wafer than depositing it by other processes.

6.3.1.1. Thermal Growth of SiO_2

The wafer is loaded into quartz furnace for silicon dioxide growth at 1100 °C to achieve a thickness of 1.0 µm following a dry-wet-dry sequence.

The process is carried out as: 5 minutes dry + 45 minutes wet +10 minutes dry at 1100 °C. Flow rate of wet O_2 is: 0.6 litres/minute.

N_2 is used as the ambient gas at a flow rate of 1 litre/minute. The combination of dry-wet-dry is used because with dry oxidation, the quality of oxide formed is high and dense but the growth rate is low. While with wet, though the quality is lower (but still good as compared to electrochemical oxidation), the growth rate is very fast. Thus, with the above combination, the time required to grow will be saved. If only dry oxidation is used, then the quality of the oxide will be, no doubt, good but lots of time will be consumed in attaining the required thickness. The wafer is ready for the next step, i.e., for LPCVD of Si_3N_4.

6.3.1.2. LPCVD of Si_3N_4

The oxidized wafer is loaded into a CVD reactor to deposit LPCVD silicon nitride at 780 °C to achieve a thickness of 0.15 µm. The combination of silicon dioxide and silicon nitride has been established as an excellent mask for KOH based etching solutions even at elevated temperatures. Though at very high temperatures and with prolong time both SiO_2 and Si_3N_4 are attacked by KOH, however, at usual temperature (up to 100 °C), they offer good resistance to attack. Also etch rate for Si_3N_4 is much slower than that of SiO_2 so a layer of Si_3N_4 of thickness 0.15 micron will prove perfect safety of the underlying substrate from being attacked by KOH.

The conditions required in the CVD chamber for LPCVD of Si_3N_4 for 0.15 microns are:

Temperature:	780 °C
Time:	38 minutes
Growth rate:	40 Å/min
Pressure:	0.4 Torr
Gases used are:	Dichlorosilane and ammonia at 20 SCCM and 200 SCCM flow rates respectively (at a ratio of 1:10)

(SCCM = standard cubic centimeter per minute)

Reaction:

$$3SiH_2Cl_2 + 10\ NH_3 \longrightarrow Si_3N_4 + 6NH_4Cl + 6H_2 \ \text{(at 780-800 °C)}$$

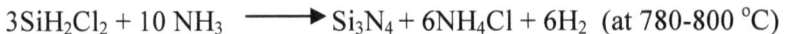

The wafer is ready for the next step, but before going further step the cleanliness must be assured. So above cleaning steps are repeated, i.e. degreasing, RCA-1 and RCA-2. Then after thorough rinsing in DI water the wafers are dried and then baked at 120 °C for about 10 minutes. This is to ensure perfect cleanliness and no moisture on the surface. Now the wafers are ready for the Photo-lithography-1, (PLG-1) for opening window.

6.3.1.3. Photolithography-1 (PLG-1)

The wafer is coated with positive photoresist (S-1813) with spinning rate at 4500 rpm for 30 seconds. Positive photoresist is used as per layout design. This gives an approximate thickness of 1.2 microns. Ambient temperature and relative humidity during photolithographic process were 24 °C and 48 % respectively.

The wafer is then pre-baked / soft-baked at 90 °C for 30 minutes. After pre-baking, the wafer is exposed to the first mask, the mask for opening window in a mask aligner with 6 seconds exposure time. After exposure the wafer is dipped in developer, MF-312 (a KOH based solution). The wafer is not directly dipped in this developer but in a solution, which is made by mixing this developer with DI water in a ratio of 1:1. In this solution the wafer is dipped for 60 seconds, then rinse in DI water and dried using N_2 gas blower. At last, wafer is observed through microscope to check for proper development of the patterns. Square window patterns can be clearly seen. Now wafer is post-baked/ hard-baked at 120 °C for 30 minutes.

Required conditions are:

Temperature:	24 °C
Relative humidity:	48 %
Positive photoresist:	S-1813
Spin rate:	4500 rpm
Thickness of PR:	~ 1.2 microns
Pre-bake:	90 °C, 30 min.
Exposure time:	6 seconds
Post-bake:	120 °C, 30 min.
Developer:	MF-312 (a KOH based solution)

After hard-baking the wafers are ready for next step, i.e. RIE of Si_3N_4 and wet etching of SiO_2 from the openings.

6.3.1.4. RIE of Si_3N_4 (0.15 Microns)

Conditions required for RIE of Si_3N_4 (0.15 microns) are:

Relative humidity:	44 %
Ambient temperature:	22 °C
Gas used is:	SF_6 in O_2 ambient at a flow rate of 150 SCCM
Flow rate ratio of SF_6 : O_2 is:	40:4
Power:	600 W
V_{dc}:	52 V
Etch pressure:	10 Pa
Etch rate:	400 Å/min
Time:	4 min. 30 sec.

Since etch rate of Si_3N_4 is 400 Å/minute, so for etching of 0.15 microns or 1500 Å the time required is about 4 minutes. But a total time of 4 minutes and 30 seconds is given so that if any residue (un-etched Si_3N_4) remains they will get etched away. O_2 is used to enhance the reaction. Instead of using SF_6, CF_4 can also be used at a flow rate of 95.4 SCCM with flow rate ratio of CF_4: O_2 = 40: 2 to obtain the same etch rate. SF_6 does not etch SiO_2 but can etch Si, while CF_4 does not etch Si but can etch SiO_2. Here SiO_2 is not to be protected from etching so either of the gases can be used.

6.3.1.5. Wet etching of SiO_2 (0.5 Microns)

SiO_2 is etched out in buffer HF. The specification of buffer HF used is:

Buffer oxide etchant (BOE) of 25:4 = 40 % NH_4F : 49 % HF volume ratio.

Etch rate at 21 °C = 830-900 Å/min.

The wafers are dipped in the BOE for 8 minutes then rinsed in DI water, dried and then checked through microscope. The color of the opening portion was greenish before etching; after 6 minutes dip it is found to be light green indicating that there is still a fine layer of SiO_2. The wafers are further given 5 minutes dip and then after rinsing in DI water and drying, they were again observed through microscope. This time the color was found to be grayish indicating that the layer is Si, meaning etching of SiO_2 is complete. After this procedure, the wafers are again given 5 minutes dip in BOE to ensure complete removal of SiO_2.

SiO_2 can also be etched out by RIE, but during wet etching the wafer can be observed through microscope to check for the complete removal and also Buffer HF does not harm Si. While RIE much depends on the operator's skill and also cost encountered is high. If the feature size is very minute then RIE should be preferred as there won't be undercutting which is most likely to occur in the case of wet etching. Now wafers are ready for stripping off of photoresist (PR).

6.3.1.6. Stripping Off of Photoresist

The samples are dipped in Methanol and boiled for 5 minutes. Then they are dipped in Acetone and boiled for 10 to 15 minutes. PR dissolved in the acetone turning it reddish is seen. Taking fresh acetone the wafers are again dipped and boiled. This step is repeated until no PR remains. This can be confirmed from the clear color of the acetone. Now wafers are ready for next step, i.e., deposition of polysilicon by LPCVD.

6.3.1.7. LPCVD of Polysilicon (Thickness 1.0 μm)

Before LPCVD of polysilicon the samples are given acetone and methanol treatment again and baked at 250 °C for two hours in oven. A complete dry surface is ensured for good quality of polysilicon deposition. In case of any traces of moisture on the wafer surface leads to foggy type of polysilicon. Polysilicon of thickness 1.0 micron is required to be deposited to act as the sacrificial layer.

The deposition conditions are:

Gas used is SiN_4 at flow rate 50 SCCM.
Operating temperature: 620 °C
Total time taken: 40 min.
Initial pressure: 0.2 Torr
Operating time pressure: 0.12 Torr
Grown Rate: ~ 250 Å/min

Reaction:

$$SiH_4 \rightarrow Si + 2H_2$$

Now samples are ready for next step, i.e. photolithography (PLG-2) for defining the sacrificial layer.

Photolithography (PLG-2) for defining the sacrificial layer is carried out using the above same procedure as for PLG-1. After hard-bake, the samples are ready for next step, i.e. poly-etch.

6.3.1.8. Etching of Polysilicon

Etching of polysilicon is done by RIE technique. Polysilicon can also be etched by wet etching technique. In this technique the wafers are dipped in poly etch solution. Poly etch is a solution for etching of polysilicon with the composition HF, HNO_3 and NH_4F. The samples are dipped in poly etch solution separately for about 5 minutes each.

Then rinse in DI water and dried using N_2 gas gun and observe through microscope.

Compositions of Poly etch: HF, HNO_3 and NH_4F.

Before going to the next step the wafers are given piranha treatment, rinsed thoroughly in DI water, dried and baked at 120 °C for 30 minutes in oven. These samples are ready for next step i.e. PECVD of (Si_3N_4 + + SiO_2 + Si_3N_4).

6.3.1.9. PECVD of (Si_3N_4 + SiO_2 + Si_3N_4).

Over the sacrificial layer of polysilicon, a composite layer of (Si_3N_4 + SiO_2 + Si_3N_4) is required for the purpose of membrane formation. Si_3N_4 exhibits greater resistance to KOH attack than SiO_2. Since front-side etching technique is to be used, lower portion as well as upper portion of SiO_2 need to be protected from KOH attack during anisotropic KOH etching of cavity. A 0.1 to 0.2 micrometer thick layer of Si_3N_4 is found to be crack free. Thicker layers of the Si_3N_4 develop strain to generate cracks while annealing at higher temperatures. An embedded layer of relatively thicker SiO_2 provides two fold advantages. One for enhancing diaphragm thickness and secondly to provide support for strained Si_3N_4 layers. It is also known that Si_3N_4 is tensile in nature while SiO_2 is compressive in nature. The combination of these two layers having opposite stresses in nature will compensate the internal stress developed with each other.

LPCVD, despite of yielding better quality, is being replaced by PECVD because PECVD is less hazardous and less time consuming. Also, after annealing at high temperature PECVD deposited layers can yield comparable quality as that of LPCVD. Annealing reduces not only the porous nature in the deposited film but also reduces strain developed within during the deposition.

6.3.1.10. PECVD of Si_3N_4 for Thickness 0.2 Microns

Operating temperature:	300 °C
Operating pressure:	0.4 Torr
Gas used:	SiH_4 (80 %), NH_3 (25 %) and N_2 (30 %)
Power:	120 Watt
Total time:	15 min.

136

Reaction:

$$3SiH_4 + 4NH_3 \longrightarrow Si_3N_4 + 12 H_2 \text{ (at 200-400 °C)}$$

6.3.1.11. PECVD of SiO₂ for Thickness 0.5 Microns

Operating temperature:	300 °C
Operating pressure:	0.3 Torr
Gas used:	Nitrous oxide, SiH_4, N_2
Power:	80 Watt
Total time:	11 min.

Reaction:

$$SiH_4 + 2N_2O \longrightarrow SiO_2 + 2N_2 + 2H_2$$

PECVD of Si_3N_4 for thickness 0.2 microns is again repeated.

The samples are then annealed at 1000 °C for one hour at N_2 ambient with flow rate of 1.0 litre/minute. Due to annealing the total thickness of the composite layer has reduced by 0.1-0.2 microns (approximately).

LPCVD of polysilicon (thickness 1.0 μm)-LPCVD of polysilicon for thickness ~1.0 μm is again repeated.

6.3.1.12. Boron Doping of Polysilicon for Piezoresistors

In order to provide piezo-effect, boron is doped using boron nitride solid sources at elevated temperature. Doping of polysilicon is carried out in order to control the sheet resistivity of the polysilicon layer and to get the piezo-effect. Thermal diffusion technique has been used for the boron doping purpose (introducing dozes above 1×10^{15} ions/cm²) in the present work.

Polysilicon is doped with boron because p-type shows better piezoresistive effect than n-type. Sheet resistivity of the doped polysilicon is an important parameter in order to decide the value of resistors. Boron diffusion can be performed up to a maximum temperature of 1200 °C but it is not advisable to go beyond 1050 °C. Beyond at this temperature the wafer boat start to fuse with the furnace tube thus rising problem while pulling out the wafer boat and using force in pulling may not only damage the wafers on the wafer boat but may also develop cracks on the furnace tube. For the boron to diffuse up to a depth of 1.0 μm (total thickness of the polysilicon layer), boron diffusion is carried out at 1050 °C for 40 minutes.

6.3.1.13. Boron Diffusion Conditions

Temperature: 1050 °C
Total time: 40 min.
Sheet resistivity: ~20.00 Ω/

For confirming the sheet resistivity measurements are done using four-point probe method. The sheet resistivity is measured in four different parts on each sample. The measured values lie within the range 19.0 to 20.50 Ω/.

Due to the doping of boron at very high temperature, Borosilicate glass (BSG) is formed on the doped polysilicon surface. To remove this BSG layer, a dielectric, the samples are first dipped in buffer HF for 2-3 minutes. This, though cannot remove the BSG layer, will loosen them. The samples are then dipped in nitric acid to get oxidized, followed by thorough rinse in DI water. Then the samples are given wet oxidation for 5 minutes, (i.e., dry 2 minutes + wet 5 minutes + dry 4 minutes) at 950 °C. After this, the samples are dipped in buffer HF to remove the BS glass. The samples are again rinsed in DI water. Here the samples show hydrophobic nature indicating no oxide layer on the surface. These samples are ready for the next step, i.e. photolithography for defining resistors. Before going to the next step the samples are baked in oven at 120 °C for about one and half hours.

Photolithography (PLG-3) for defining resistors is carried out using the above same procedures as for PLG-1and PLG-2.

Etching of boron doped polysilicon for defining resistors is done by reactive ion etching technique.

6.3.1.14. Metallization for Contact Lines and Contact Pads

Cleanliness of the samples is an important necessity for metallization. So before going for metallization the samples are given degreasing, dried and then baked at 120 °C for 30 minutes. These samples are now ready for metallization.

Aluminium is the metal used and metallization is carried out by RF magnetron sputtering. Here the target electrode/plate is the aluminium and the noble gas used is Argon. The required Aluminium thickness is 1.0 micron.

The required conditions are:

Technique used:	RF magnetron sputtering
Target electrode:	Al
RF frequency:	13.56 MHz
Power:	500 Watts
Initial pressure before Ar:	5×10^{-6} mTorr
Pressure after introducing Ar	$\sim 3 \times 10^{-3}$ mTorr
Total time taken for 1.0 micron:	1 hour

Metallization can also be done for other metals such as Au, Ni, Pt, and W etc. by e-beam evaporation method. Varian's e-beam metallization unit at CEERI laboratory, Pilani is shown in Figure 6.5.

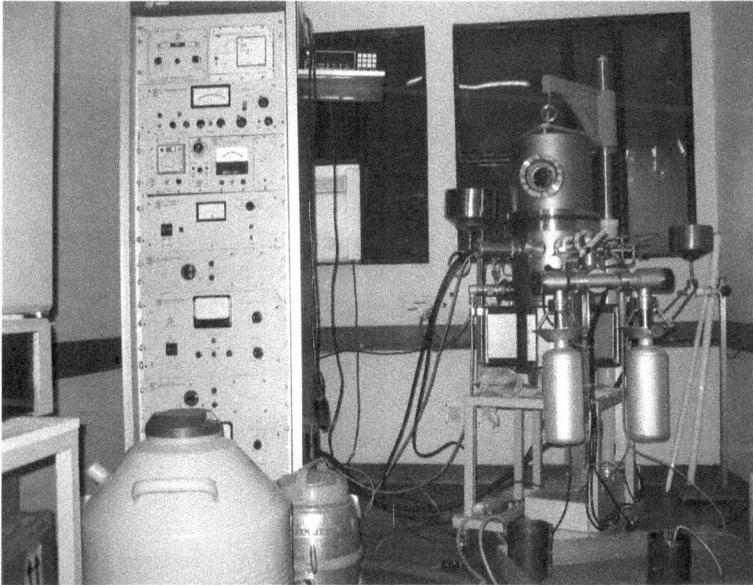

Figure 6.5. Varian's e-beam metallization unit at CEERI laboratory, Pilani.

The samples are ready for the next step i.e. PLG for defining metal contact line and contact pads.

6.3.1.15. Photolithography (PLG-4) for Defining Metal Contact Lines and Contact Pads

PLG –4 is carried out using the above same procedures as for PLG-1, 2 and 3. After hard baking of 30 minutes the samples are ready for Aluminium etching.

6.3.1.16. Aluminium Etching

Aluminium etching for defining the contact lines and contact pads is carried out by wet etching. Aluminium etchant is the solution used. This solution is prepared in the laboratory by using the following compositions:

Aluminium etchant:	$N_3PO_4 : HNO_3 : CH_3COOH : H_2O = 4 : 1 : 4 : 1$
Etch rate:	1.0 micron/30 minutes

After Al etching the samples are thoroughly rinsed in DI water, dried using N_2 gun and observed through microscope. The metal contact lines and pads are well defined.

6.3.1.17. Sintering

Sintering is a method of increasing adhesion between particles as they are heated. This step is required for ensuring proper ohmic contact of the metal deposited with the underlying layer.

The conditions required are:

Temperature:	450 °C in forming gas ambient
Time:	30 minutes

Forming gas, which is used as an ambient, consists of 5-10 % Hydrogen and the rest Nitrogen. The presence of Hydrogen improves the quality.

APCVD of SiO$_2$ (~0. 5 μ) and PECVD of Si$_3$N$_4$ (~0. 3 μ) are carried out for the purpose of protection of contact lines and resistor pads due to the risk of KOH etching during cavity formation. After that the samples are annealed at 500 °C with N$_2$ gas for one hour to densify the layers deposited.

Photolithography-5 (PLG-5) is done using the same procedures as in PLG-1, 2, 3 and 4, to open etch holes for the purpose of cavity formation. Since the feature size of mask used during PLG-5 is quite small, much care must be taken during aligning. Also care must be taken to ensure perfect protection of the diaphragm so thick photoresist (AZ P4903) is used in this photolithography.

6.3.1.18. Details of the PLG-5

Positive photoresist used:	AZ P4903
Spin rate:	5500 rpm for 30 seconds
Exposure time:	10 seconds
Developer used:	AZ400K at 1:1 with DI water
Thickness of the resist:	7.2 microns (measured by Talystep)

RIE (Dry etching) is carried out for etching of composite layer of (SiO$_2$ + Si$_3$N$_4$) having thickness (~ 1.7 μ). Now wafers are ready for KOH etching.

6.3.1.19. KOH Etching

It is done to etch the polysilicon layer (sacrificial layer) as well as silicon material in the bulk of substrate for the formation of V-shape cavity under the membrane. Experimental set-up for KOH etching is shown in Figure 6.6. The reflux condenser used on top of the beaker maintains the concentration of KOH in the solution by condensing the vapor back to solution.

Before anisotropic etching, the samples are given degreasing treatment and 5 % HF dip to ensure complete removal of native oxide followed by thorough rinse in DI water otherwise the presence of native oxide reduces the etch rate. After maintaining the temperature of KOH solution at ~80 °C, KOH etching is carried out on the samples.

Figure 6.6. A photograph of KOH set-up
(At MEMS laboratory, CEERI, Pilani).

Since the opening of the square cavity is 100 μm × 100 μm and the wafer orientation of the samples are Si<100>, the depth of the V-shaped cavity that will be formed by anisotropic etching can be found out as given in Figure 6.7.

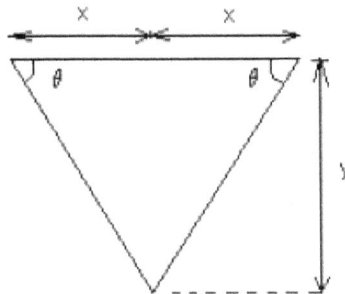

Figure 6.7. Geometical view of conical V- shaped etched cavity.

Here, 'x' is 100/2, i.e., 50 μm and the angle θ (the angle between <100> and <111> plane) is 54.74°. Thus, 'y' i.e. the depth of the V-shaped cavity that will be formed by anisotropic etching will be

$$
\begin{aligned}
Y &= x \tan \theta \\
&= 50 \tan 54.74° \\
&= 50 \times 1.41 \; \mu m \\
&= 70.72 \; \mu m
\end{aligned}
$$

Since the etch rate of KOH for Si at 80 °C is 1.2 μm/minute, KOH etching of the samples for the formation of V- shaped cavity of depth ~ 70.72 μm takes roughly one and half-hours. So after one and half hours, the samples are taken out, rinsed thoroughly in DI water and dried using N_2 gas gun. Then the samples are observed through a powerful microscope to check for the completion of etching. Recalling that the diaphragm consists of a composite layer of Si_3N_4 + SiO_2 + + Si_3N_4, this composite layer is transparent so the cavity formed beneath this layer can be easily seen through it. From the observation it is found that much of the etching still remains. This shows that the etch rate here is much slower than it normally is. Also it is found that etching starts from the two sides where the etch holes are opened as expected. This slow etch rate is due to the passage of very little amount of KOH etchant to the sacrificial layer and then to the silicon, where it is opened to be etched, is quite narrow. So etching is continued further for half an hour more. This duration is given after approximate calculation from the observed condition with sufficient additional time to ensure complete etching. Additional time will not harm the samples as the etching will be stopped by itself as soon as the V-shaped cavity (coneshaped) is formed. This is the advantage of anisotropic etching of Si<100>. After this, the samples are rinsed thoroughly in DI water, dried using N_2 gas gun and observed through the microscope. This time proper formation of V-shaped cavity can be clearly seen, thus ensuring complete etching.

The next step, as per planned before, is **APCVD of SiO₂ and PECVD of Si₃N₄** for the purpose of sealing of cavity and passivation.

Photolithography-6 (PLG-6) and RIE is done using above the same procedures as previously done for opening of metal contact pads.

At last, **PLG-7** is done for grid formation and dicing.

6.3.2. Detail of Process Sequence (II)

This process is different from process sequence (I) by some steps. Here cavity is formed at initial stage while in process (I) at final stage. Some modifications made in this process are as follows:

6.3.2.1. Membrane and Cavity Formation

In this process membrane is formed by LPCVD (Si_3N_4 + SiO_2 + Si_3N_4), then PLG-3 and RIE is done to etch composite layer of (Si_3N_4 + SiO_2 + + Si_3N_4) to open holes for the passage of KOH etchant through which etchant enters in the bulk of substrate. After that KOH etching is done for the formation of V- shaped cavity.

6.3.2.2. Sealing of the Cavity and Deposition of Resistors

Here the cavity is sealed by LPCVD of polysilicon and this deposited polysilicon is doped with boron for the purpose of resistors formation, so no extra deposition is required for sealing of cavity. Passivation of the device is done only by PECVD of Si_3N_4.

Thus from the technological point of view the process sequence (II) is more better then process sequence (I).

References

[1]. J. Akhtar, B. B. Dixit, B. D. Pant, V. P. Deshwal, and B. C. Joshi, A process to control diaphragm thickness with a provision for back to front alignment in the fabrication polysilicon piezoresistive pressure sensor, *Sensor Review*, Vol. 23, No. 4, 2003, pp. 311- 315.
[2]. H. Guckel, Surface micromachined pressure transducers, *Sensors and Actuators A*, 28, 1991, pp. 133-146.
[3]. K. E. Peterson, Silicon as a mechanical material, in *Proceedings of the IEEE*, 70, 5 1982, pp. 420-457.
[4]. H. Guckel, D. Burns, Planar processed polysilicon sealed cavities for pressure transducers array, *IEDM*, 1984, pp. 223.
[5]. S. Sugiyama, K. Shimaoka, O. Tabata, Surface micro-machined micro diaphragm pressure sensors, in *Proceedings of the 6th Int. Conf. on Solid State Sensors and Actuators, Transducers*, 1991, pp. 188.

[6]. E. Bassous, Fabrication of novel three dimensional microstructures by anisotropic etching of (100) and (110) silicon, *IEEE Trans. on Electron Devices,* ED 25, 1978, pp. 1178.

[7]. S. Suwazono, H. Tanigawa, M. Hirata, Diaphragm thickness control sin silicon pressure sensor using an anodic oxidation etch-stop, *J. of Electromechanical Society,* 134, 1987, pp. 2037-2041.

[8]. Wen Zhiyu, Wu Ying, Zhang Zhengyuan, Xu Shilu, Huang Shanglian, Li Youli, Development of an integrated vacuum microelectronic tactile sensor array, *Sensors and Actuators A*, 103, 2003, pp. 301-306.

[9]. R. M. Tiggelaar, P. Van Male, J. W. Berenschot, J. G. E. Gardeniers, R. E. Oosterbroek, M. H. J. M. de Croon, J. C. Schouten, A. van den Berg, M. C. Elwenspoek, Fabrication of a high-temperature microreactor with integrated heater and sensor patterns on an ultrathin silicon membrane, *Sensors and Actuators A: Physical,* Vol. 119, Issue 1, 28 March, 2005, pp. 196-205.

[10]. R. Kressmann, M. Klaiber, and G. Hess, Silicon condenser microphones with corrugated silicon oxide/nitride electret membranes, *Sensors and Actuators A*, 100, 2002, pp. 301-309.

[11]. Peng Chih-Tang, Lin Ji-Cheng, Lin Chun-Te, Chiang Kuo-Ning, Performance and package effect of a novel piezoresistive pressure sensor fabricated by front side etching technology, *Sensors and Actuators A: Physical,* Vol. 119, Issue 1, 28 March, 2005, pp. 28-37.

[12]. J. Akhtar, B. B. Dixit, B. D. Pant, V. P. Deshwal, B. C. Joshi, P. R. Deshmukh, Kamaljit Rangra, O. P. Wadhawan, and S. Ahmad, Effect of Mask edgealignment on anisotropic etching of concave and convex geometries of (100) silicon in aqueous KOH solution, *Indo-Japanese Workshop on Micro System Technology*, University of Delhi, Delhi, India, November 23-25, 2000.

Chapter 7

In-Process Observations

In this chapter the photographs observed at every stage during the fabrication process (I) and (II) are shown and analyzed. The photographs of process sequence (I) and (II) taken with the help of optical inspection microscope are as under:

7.1. Photographs for Process Sequence (I)

7.1.1. Photograph after PLG-1, SiO$_2$ and Si$_3$N$_4$ Etching

The square shaped window having size of 100 μm × 100 μm is opened for the purpose of membrane formation. The light greenish color in the square shows opened window indicating silicon material and the remaining portion is covered by masking layers of silicon dioxide and silicon nitride (Figure 7.1).

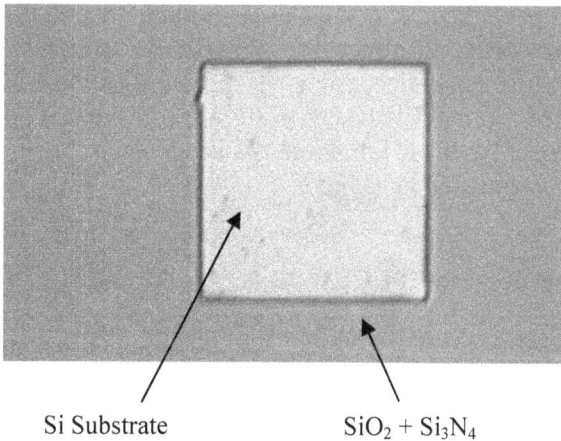

Si Substrate   SiO$_2$ + Si$_3$N$_4$

Figure 7.1. Photograph taken after PLG-1, SiO$_2$ and Si$_3$N$_4$ etching (for opening square window of size 100 μm × 100 μm).

7.1.2. Photograph after PLG-2 and Poly Etch (with Photoresist)

The sacrificial layer of polysilicon with photoresist defined after PLG-2 and poly etch is shown in Figure 7.2.

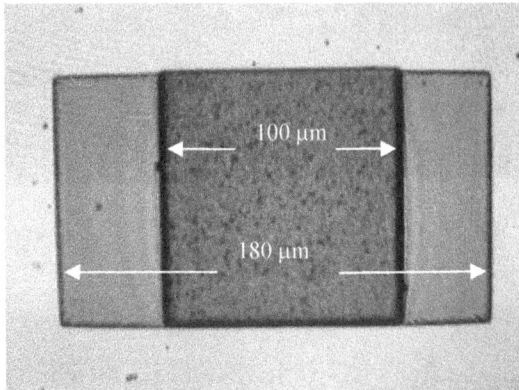

Figure 7.2. Shows the sacrificial layer of polysilicon with photoresist defined after PLG-2 and poly etch (Size of sacrificial layer = 180 μm × 100 μm).

7.1.3. Photograph after PLG-2 and Poly Etch (without Photoresist)

The LPCVD polysilicon film deposited on the both sides of the silicon wafer produce buckling effect due to excessive compressive stresses developed in the polysilicon film. However, buckling effect has not been observed in polysilicon at this stage of fabrication. The side spread of the polysilicon film over the square window provides etch channels for lateral etching (Figure 7.3).

7.1.4. Photograph after PECVD (Si_3N_4 + SiO_2 + Si_3N_4)

When wafer is annealed at 1000 °C with N_2 ambient for one hour after the step (6) in process sequence (I), cracks are observed in the thin composite layer of PECVD (Si_3N_4 + SiO_2 + Si_3N_4) as shown in Figure 7.4.These cracks are perhaps due to the stresses in both the layer in an opposite manner i.e. stresses in silicon dioxide are more compressive and tensile in silicon nitride. Here it is suggested, therefore, to form the membrane of composite layer of LPCVD ($Si_3N_4 + SiO_2 + Si_3N_4$).

Figure 7.3. Shows polysilicon sacrificial layer defined after stripping off photoresist.

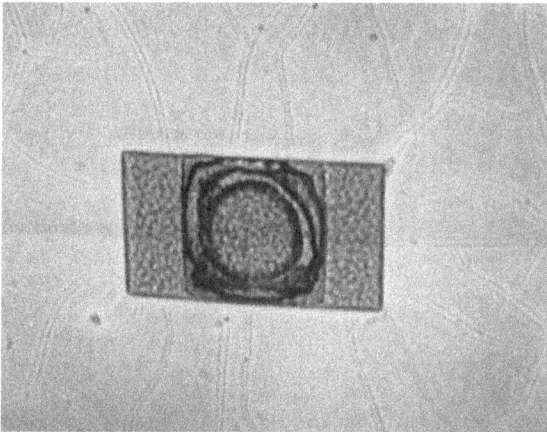

Figure 7.4. Shows the cracks observed in the PECVD ($Si_3N_4 + SiO_2 + Si_3N_4$) after annealing at $1000\,^0C$.

7.1.5. Photograph after (LPCVD of Polysilicon + Boron Doping + + PLG-3 + RIE of Polysilicon) Defining of Boron Doped Polysilicon Resistors

In the Figure 7.5, four resistors (piezoresistors) of boron-doped polysilicon are shown. These four resistors are fabricated for the purpose of half Wheatstone bridge configuration in such a way two resistors are on the membrane and rest two outside the square cavity region.

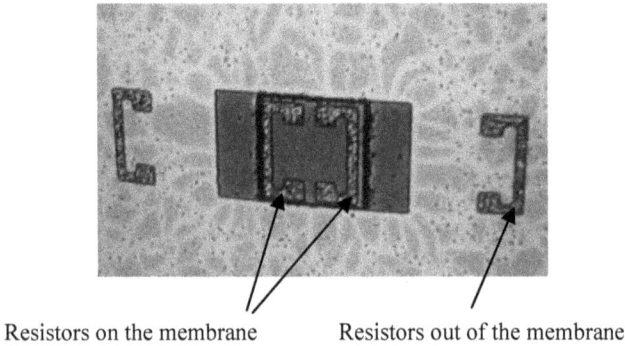

Resistors on the membrane   Resistors out of the membrane

Figure 7.5. Shows the resistors deposited of boron doped polysilicon.

7.1.6. Photograph after (Al metallization + PLG-4 + Wet Etching of Al) Defining Metal Contact Pads and Contact Lines

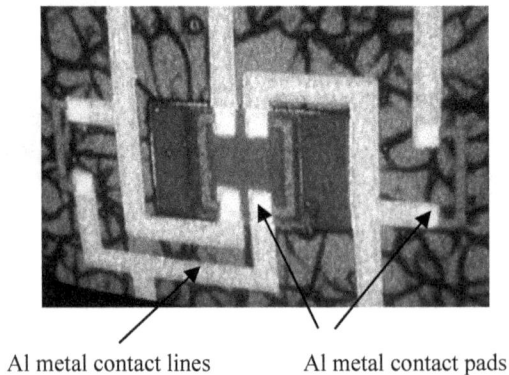

Al metal contact lines   Al metal contact pads

Figure 7.6. Metallic contact lines and contact pads.

7.1.7. Photographs after KOH Etching for Cavity Formation

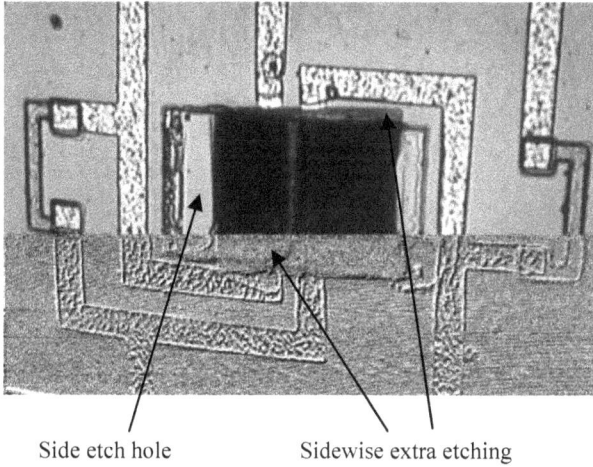

Side etch hole   Sidewise extra etching

Figure 7.7. Photograph taken after V-shaped cavity formation.

In Figures 7.8, 7.9 and 7.10, it can be seen clearly KOH etching starts from the side holes through channels opened, thus a barrier remain in middle indicating incomplete etching which requires extra time. The V-shaped cavity formation observed as focused through the transparent diaphragm is shown in Figure 7.8. The photographs are also taken, shown in Figures 7.9 and 7.10, from left and right side during KOH etching for cavity formation.

Figure 7.7 shows extra sidewise KOH etching due to misalignment of mask edge. In case of misaligned mask edges, concave geometries increase in size while convex geometries are reduced. Due to mask edge-misalignment, the etched cavity becomes wider during KOH operation resulting into damages to the support of the membrane and therefore metallic interconnection lines as shown in Figure 7.11. With the decreasing dimensions of the cavities, mask edge alignment becomes more and more sensitive, failing which the ultimate devices would be of no use, giving rise to a zero yield. A 2″ diameter silicon wafer has been processed to its final stage giving zero percent yields with this misalignment [1-3].

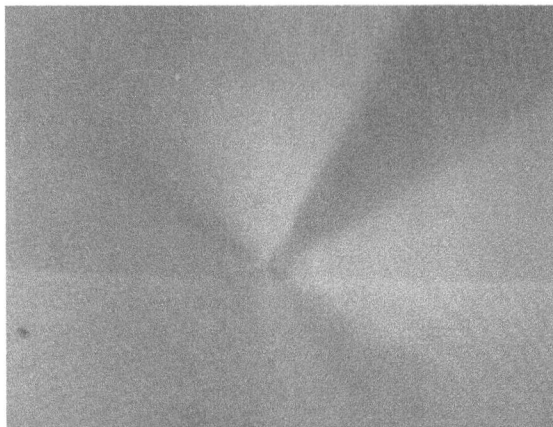

Figure 7.8. The V-shaped cavity formation observed
as focused through the transparent diaphragm.

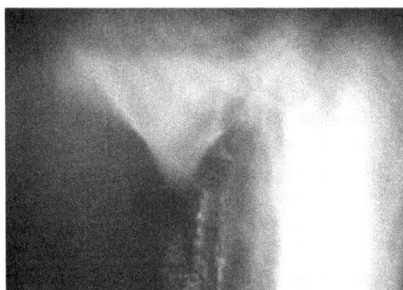

Figure 7.9. From left side.

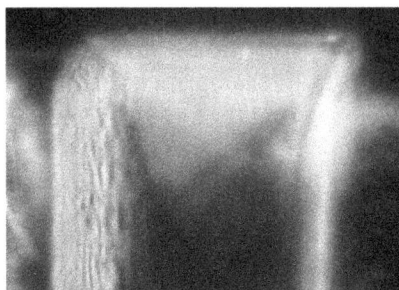

Figure 7.10. From right side.

As discussed above, due to misaligned mask edges during KOH
operation for cavity formation, most of the membranes are found crack
and broken, as shown in Figure 7.12, and in remaining devices left
metallic lines are found disconnect from the resistor pads as shown in
Figure 7.11. Thus all the devices after KOH operation (cavity
formation) are found of no use. Then process sequence (II) is started in
which cavity is formed at initial stage. The photographs taken for
process sequence (II) are as under.

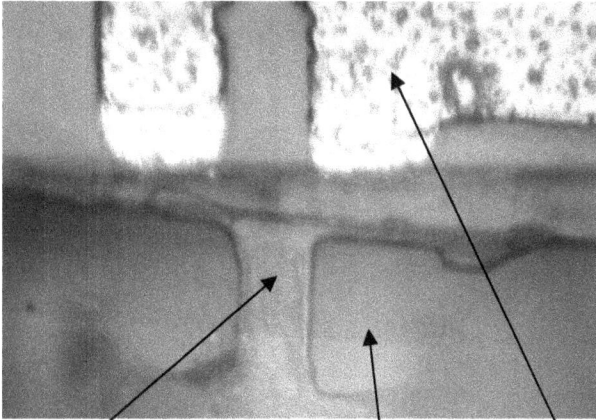

Disconnectivity of contact lies   Resistor's pad   Contact line
from resistor's pad   over membrane

Figure 7.11. Shows disconnected metallic lines from the resistors deposited over membrane.

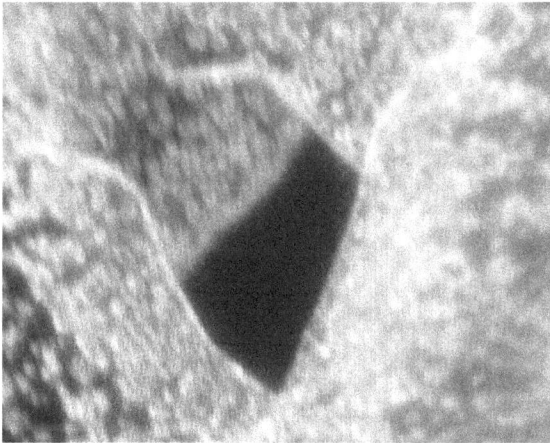

Figure 7.12. Shows a broken membrane.

7.2. Photographs for Process Sequence (II)

7.2.1. Photograph after LPCVD (Si_3N_4 + SiO_2 + Si_3N_4)

In the Figure 7.13, photograph (after step 6 in process sequence II) of LPCVD composite layer of silicon nitride and silicon dioxide is shown. In this composite layer no cracks have been observed and quality of deposited film is also better than composite layer of PECVD (Si_3N_4 + SiO_2 + Si_3N_4) in process sequence (I).

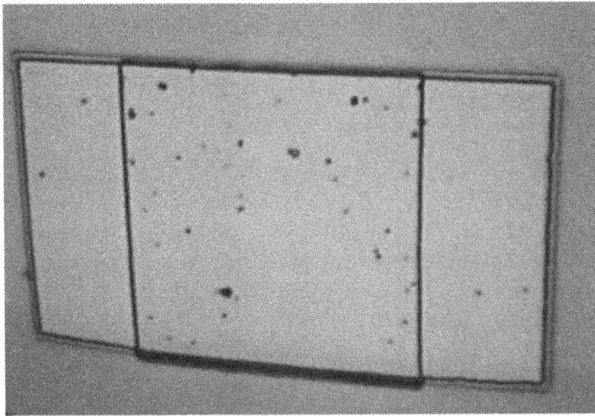

Figure 7.13. LPCVD composite layer of (Si_3N_4 + SiO_2 + Si_3N_4).

7.2.2. Photographs after PLG-3 and RIE of LPCVD (Si_3N_4 + SiO_2 + Si_3N_4)

In Figures 7.14 and 7.15, opened etch holes having size of 20 μm × 100 μm are shown. These holes are opened for the purpose of cavity formation through which KOH etchant enters in the bulk of silicon substrate.

7.2.3. Photographs after KOH Operation

Figure 7.16 shows KOH etching starts from the side holes through channels opened, thus a barrier remain in middle indicating incomplete etching which requires extra time. As etching time is further increased the barrier in between becomes smaller and only a point remains in the center as shown in Figure 7.17.

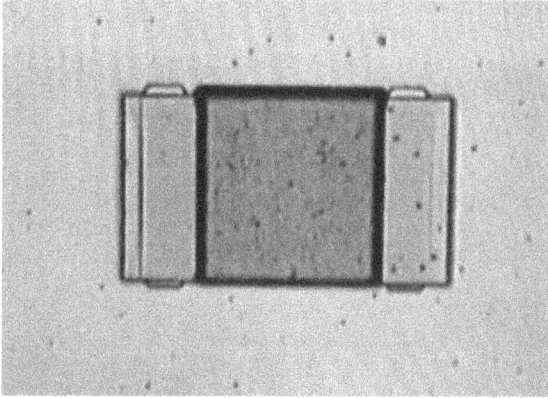

Figure 7.14. Shows patterning of etch holes by PLG-3.

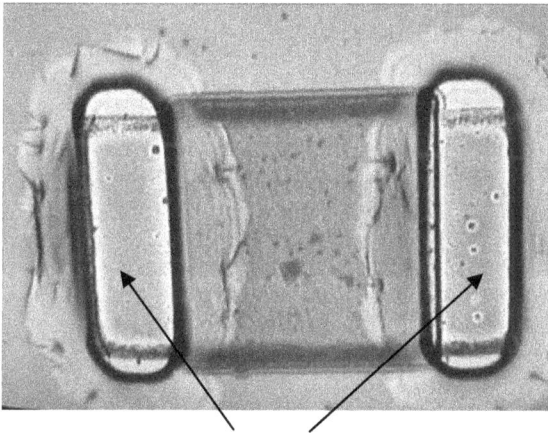

Opened etch holes

Figure 7.15. Shows opening of etch holes after RIE of composite layer.

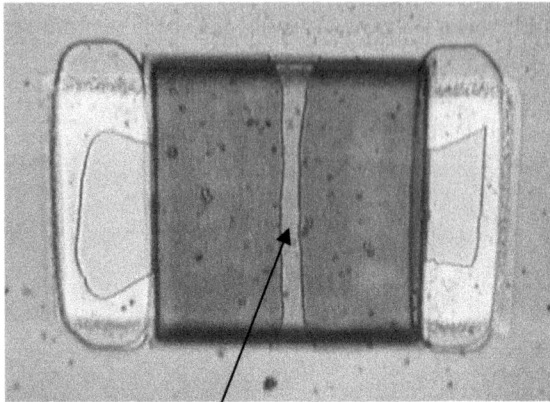

Barrier left in middle

Figure 7.16. Shows photograph taken through transparent membrane during KOH etching.

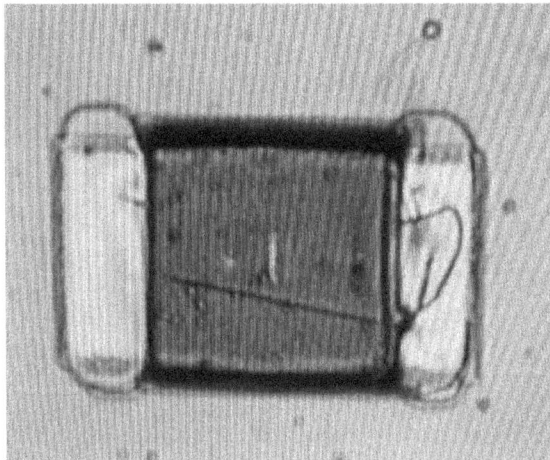

Figure 7.17. Shows central portion remaining after complete KOH etching.

After sealing the cavity by LPCVD polysilicon and diffusion of boron in it, the photograph is shown in Figure 7.18. From the photograph it is seen that the etched holes are sealed. Careful observation reveals that the etch holes are completely sealed and due to the formation of vacuum inside the diaphragm bends inward.

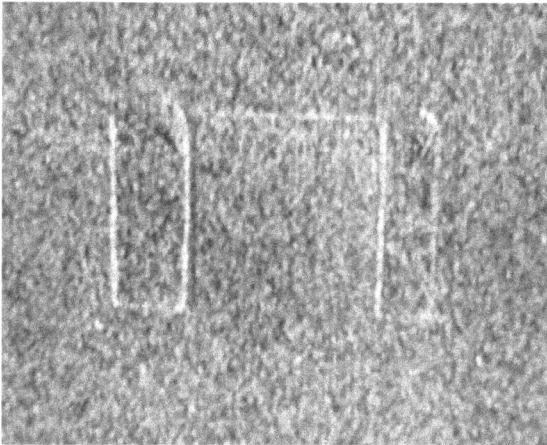

Figure 7.18. Shows etch hole sealed by LPCVD of polysilicon.

7.2.4. Photographs after Defining Resistors and Metallic Lines

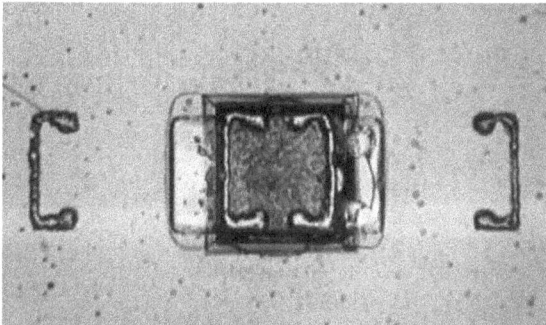

Figure 7.19 Shows deposited boron doped polysilicon piezoresistors.

157

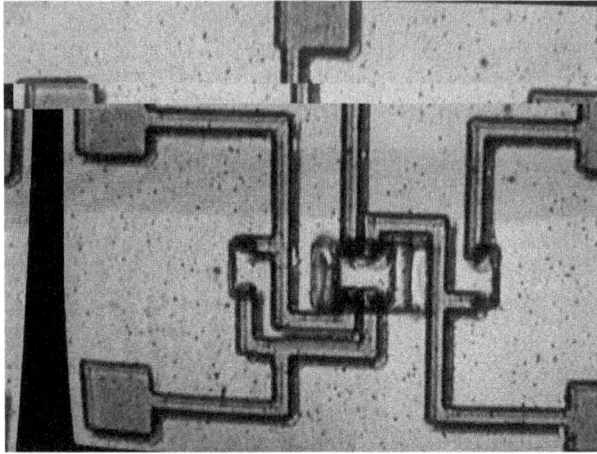

Figure 7.20. Shows the defined metallic lines after PLG-5.

Figure 7.21. Shows the defined metallic lines after Aluminium etching
and stripping out photoresist.

Thus after passivation the process sequence (II) is completed.

The bonding, packaging and a rigorous characterization of the fabricated device is in progress and the results related to its testing in actual environment shall be reported at a later stage.

The values of boron-doped polysilicon piezoresistors deposited outside the diaphragm (i.e. R_1 and R_3) in the Wheatstone bridge are measured randomly on the wafer by probe method, as shown in Figures 7.22 and 7.23.

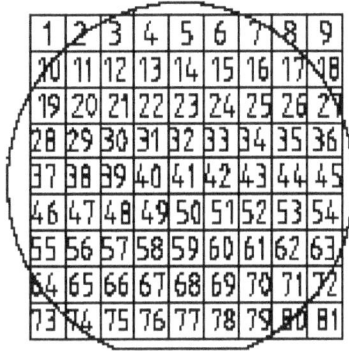

Figure 7.22. Wafer map.

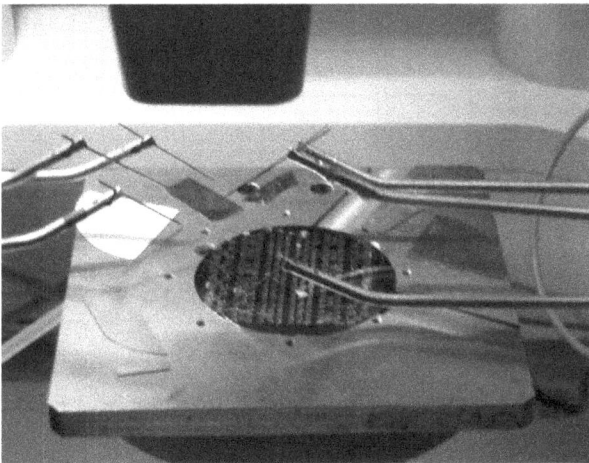

Figure 7.23. Resistors measurement system at CEERI, Pilani.

Values of the resistances are measured and compared for a relative deviation from the mean value of the two resistors deposited outside the diaphragm in the bridge on a chip. The data recorded on eighteen devices is listed in Table 7.1. The typical variation is in the range of 3 to 15 %.

Table 7.1. Measured values of polysilicon piezoresistors (as fabricated) outside the diaphragm.

Device No.	R_1 (kΩ)	R_3 (kΩ)	R (Mean) kΩ	Variation of resistances kΩ	% Variation
1.	0.26	0.29	0.27	0.03	-3.84, +6.89
2.	0.29	0.32	0.30	0.03	-3.44, +6.25
3.	0.28	0.33	0.30	0.05	-7.14, +9.09
4.	0.27	0.32	0.29	0.05	-7.40, +9.37
5.	0.29	0.30	0.29	0.01	0.00, +3.33
6.	0.26	0.29	0.27	0.03	-3.84, +6.89
7.	0.30	0.29	0.29	0.01	0.00, +3.33
8.	0.30	0.28	0.29	0.02	-3.57, +3.33
9.	0.29	0.28	0.28	0.01	0.00, +3.44
10.	0.29	0.29	0.29	0.00	0.00, 0.00
11.	0.31	0.33	0.32	0.02	-3.22, +3.03
12.	0.24	0.30	0.27	0.06	-12.5, +10.0
13.	0.32	0.31	0.31	0.01	0.00, +3.12
14.	0.26	0.31	0.28	0.05	-7.69, +9.67
15.	0.28	0.35	0.32	0.05	-14.28, +8.57
16.	0.28	0.30	0.29	0.02	-3.57, +3.33
17.	0.23	0.31	0.27	0.08	-17.39, +12.90
18.	0.30	0.31	0.30	0.01	0.00, +3.33

Note: % variation shows the variation of resistances (higher and lower values) with respect to the mean value.

References

[1]. J. Akhtar, B. B. Dixit, B. D. Pant, V. P. Deshwal, B. C. Joshi, P. R. Deshmukh, Kamaljit Rangra, O. P. Wadhawan, and S. Ahmad, Effect of Mask edge-alignment on anisotropic etching of concave and convex geometries of (100) silicon in aqueous KOH solution, *Indo-*

Japanese Workshop on Micro System Technology, University of Delhi, Delhi, India, November 23-25, 2000.

[2]. P. A. Alvi, S. K. Gupta, K. M. Lal, and J. Akhtar, Yield degradation due to mask edge-misalignment in the fabrication of micro pressure sensor using front- side lateral etching technology, *The 1st Indo-US Workshop on Spectroscopy; Future Trends in Spectroscopy: Application to National Security,* Department of Physics, Banaras Hindu University, Varanasi (U.P.), India, Jan 9-11, 2006.

[3]. P. A. Alvi, J. Akhtar, K. M. Lal, S. A. H. Naqvi, and A. Azam, Design and Fabrication of Micromachined Absolute Micro Pressure Sensor, *Sensors & Transducers*, Vol. 96, Issue 9, September 2008, pp.1-7.

Chapter 8

Summary

The work carried out on front side lateral etching technology is reported in this book. This study proposes a novel front side etching fabrication process for silicon based piezoresistive pressure sensor to replace the conventional back-side bulk micro-machining. This novel structure pressure sensor can achieve the distinguishing features of the chip size reduction and fabrication costs degradation.

A process for the fabrication of a vacuum sealed cavity in (100) silicon with the piezoresistors configured in half Wheatstone bridge has been developed by integrating conventional microelectronic processes with MEMS technology using front side lateral etching technique. The sealing of the cavity provides a way to set the desired range of the pressures to be measured. The technology is compatible to fabricate tiny cavities of micron size for a number of applications as bio sensors. The feasibility of the process has been shown by delineating polysilicon resistors of desired values over the cavity. The polysilicon resistors can be realized technologically by controlling the sheet resistivity of the polysilicon layer using doping of requisite species. Use of polysilicon as a sacrificial layer is an extra advantage while doing anisotropic etching of (100) silicon under the membrane.

The process has novelty in making thin membranes of external layers in silicon while maintaining advantages of bulk micromachining. Size of the cavity can be reduced to the limits of photolithography used in the process.

On the basis of experimental studies carried out and results obtained from them following conclusions can be drawn:

- Silicon dioxide is a good insulator having band gap of 8 eV, and therefore can effectively separate different layers of conducting materials with a little electrical interference. Hence, it can be used as a structural material in the MEMS devices.

163

- Silicon nitride, having band gap 40 % smaller than that of silicon dioxide, provides significantly less electrical isolation than that of silicon dioxide.

- The grain size of the polysilicon film has been observed invariant with varying temperatures during boron doping. The heat energy supplied due to increasing temperature results into dependent orderly arrangement of the grains with no change in the grain size that provides a new front in the area of nanotechnology.

- Polysilicon material can be used as a sacrificial layer and as well as for the formation of piezo-resistors by doping of suitable species e.g. boron or phosphorous.

- Use of polysilicon film as a sacrificial layer exemplifies the importance of advanced micro-electro-mechanical (MEMS) devices such as the fabricated device in the present work.

- Aqueous KOH wet etchant is found suitable for anisotropic etching in order to make V-shaped conical cavity into Si (100) substrate. Because it shows highly anisotropic behavior; etching <111> Si planes at a much slower rate than other planes.

- The etch rate variation in Si (100) with concentration of KOH is less pronounced at lower temperatures like 50 °C and 60 °C than at 70 °C.

- Higher KOH concentrations improve etched surface finish.

- Etch rate for a particular concentration increases with temperature.

- Thin membrane offers maximum stress and hence maximum sensitivity for the pressures applied.

- The maximum allowable pressure increases with increasing membrane thickness.

- The variation in maximum allowable pressure with thickness of membrane is less pronounced for the larger area of membrane.

- The developed fabrication technique for micro pressure sensor is capable to provide nearly 1089 chips (i.e. micro pressure sensor) of 1 mm × 1 mm size on a two inch diameter Si (100) wafer.

- The chip size reduction of the micro pressure sensor could elevate the sensor chip counts per wafer and thereby reduce the manufacturing cost.

- This chip design is easy to manufacture.

- This pressure sensor could be used in space, where pressure is less than atmospheric pressure; also it could be used as a biosensor.

- The range of pressures applied (i.e. greater or less than atmospheric pressure) can be decided by methodology of sealing of tiny cavity.

- The most crucial factors are the controls of the sheet resistivity of the polysilicon layer and of the dimensions of the diaphragm during bulk- micromachining of Si (100).

- The sheet resistivity of polysilicon layer can be controlled by layer thickness, type of dopant and the temperature at which dopant is diffused.

- The design part largely depends on the simulated results of the stress profile on a strained diaphragm, which helps in the placements of resistors and also in the optimization of their shapes.

- Front side lateral etching technique is better than front side normal etching technique because the later technique is based on etching via hole from the front side of the silicon wafer and the presence of holes on the membrane further limits the applications of the proposed technique.

- The lateral etching rate of the final V- groove formation has been found seven times slower than normal etching of the square window of the same size with rest of similar parameters.

- With the decreasing dimensions of the cavities, mask edge alignment becomes more and more sensitive failing which the ultimate devices would be of no use, giving rise to a zero yield.

- Cracks are observed in the thin composite layer of (SiO_2 + Si_3N_4) deposited by PECVD, to form thin membrane, over the sacrificial layer of polysilicon on high temperature processing i.e. annealing at $1000\ °C$; therefore it is suggested to form the membrane of LPCVD (SiO_2 + Si_3N_4).

- The KOH etching of Si is hazardous to the microelectronic processes and therefore is suggested to carryout the bulk micromachining in the last.

- The fabrication process developed is easy and feasible for micro sensor technologies.

Future Scope of the Work

The developed absolute micro pressure sensor could be used in order to measure pressure of both the ranges i.e. greater or less than atmospheric pressure. The range of pressures applied can be decided by the methodology of sealing of tiny cavity. However, the bonding, packaging and a rigorous characterization of the fabricated device is essential before its commercial use, which is in progress and the results related to its testing in actual environment shall be reported at a later stage.

Index

169

International Frequency Sensor Association (IFSA) Publishing

ADVANCES IN SENSORS:
REVIEWS **1**

Modern Sensors, Transducers and Sensor Networks

Sergey Y. Yurish, Editor

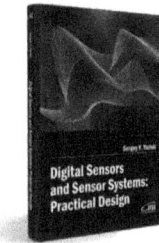

Modern Sensors, Transducers and Sensor Networks is the first book from the Advances in Sensors: Reviews book Series contains dozen collected sensor related state-of-the-art reviews written by 31 internationally recognized experts from academia and industry.

Built upon the series Advances in Sensors: Reviews - a premier sensor review source, the *Modern Sensors, Transducers and Sensor Networks* presents an overview of highlights in the field. Coverage includes current developments in sensing nanomaterials, technologies, MEMS sensor design, synthesis, modeling and applications of sensors, transducers and wireless sensor networks, signal detection and advanced signal processing, as well as new sensing principles and methods of measurements.

Formats: printable pdf (Acrobat) and print (hardcover), 422 pages

ISBN: 978-84-615-9613-3,
e-ISBN: 978-84-615-9012-4

Modern Sensors, Transducers and Sensor Networks is intended for anyone who wants to cover a comprehensive range of topics in the field of sensors paradigms and developments. It provides guidance for technology solution developers from academia, research institutions, and industry, providing them with a broader perspective of sensor science and industry.

http://sensorsportal.com/HTML/BOOKSTORE/Advance_in_Sensors.htm

International Frequency Sensor Association (IFSA) Publishing

Digital Sensors and Sensor Systems: Practical Design

Sergey Y. Yurish

The goal of this book is to help the practicians achieve the best metrological and technical performances of digital sensors and sensor systems at low cost, and significantly to reduce time-to-market. It should be also useful for students, lectures and professors to provide a solid background of the novel concepts and design approach.

Book features include:

● Each of chapter can be used independently and contains its own detailed list of references
● Easy-to-repeat experiments
● Practical orientation
● Dozens examples of various complete sensors and sensor systems for physical and chemical, electrical and non-electrical values
● Detailed description of technology driven and coming alternative to the ADC a frequency (time)-to-digital conversion

Formats: printable pdf (Acrobat) and print (hardcover), 419 pages

ISBN: 978-84-616-0652-8,
e-ISBN: 978-84-615-6957-1

Digital Sensors and Sensor Systems: Practical Design will greatly benefit undergraduate and at PhD students, engineers, scientists and researchers in both industry and academia. It is especially suited as a reference guide for practicians, working for Original Equipment Manufacturers (OEM) electronics market (electronics/hardware), sensor industry, and using commercial-off-the-shelf components

http://sensorsportal.com/HTML/BOOKSTORE/Digital_Sensors.htm